Florian A. Werner

Effects of Human Disturbance on Tropical Montane Epiphyte Assemblages

Florian A. Werner

Effects of Human Disturbance on Tropical Montane Epiphyte Assemblages

Studies from the Ecuadorian Andes

Südwestdeutscher Verlag für Hochschulschriften

Impressum/Imprint (nur für Deutschland/ only for Germany)
Bibliografische Information der Deutschen Nationalbibliothek: Die Deutsche Nationalbibliothek verzeichnet diese Publikation in der Deutschen Nationalbibliografie; detaillierte bibliografische Daten sind im Internet über http://dnb.d-nb.de abrufbar.

Alle in diesem Buch genannten Marken und Produktnamen unterliegen warenzeichen-, marken- oder patentrechtlichem Schutz bzw. sind Warenzeichen oder eingetragene Warenzeichen der jeweiligen Inhaber. Die Wiedergabe von Marken, Produktnamen, Gebrauchsnamen, Handelsnamen, Warenbezeichnungen u.s.w. in diesem Werk berechtigt auch ohne besondere Kennzeichnung nicht zu der Annahme, dass solche Namen im Sinne der Warenzeichen- und Markenschutzgesetzgebung als frei zu betrachten wären und daher von jedermann benutzt werden dürften.

Verlag: Südwestdeutscher Verlag für Hochschulschriften Aktiengesellschaft & Co. KG
Dudweiler Landstr. 99, 66123 Saarbrücken, Deutschland
Telefon +49 681 37 20 271-1, Telefax +49 681 37 20 271-0
Email: info@svh-verlag.de
Zugl.: Göttingen, Georg-August-Universität, Dissertation, 2008

Herstellung in Deutschland:
Schaltungsdienst Lange o.H.G., Berlin
Books on Demand GmbH, Norderstedt
Reha GmbH, Saarbrücken
Amazon Distribution GmbH, Leipzig
ISBN: 978-3-8381-1646-4

Imprint (only for USA, GB)
Bibliographic information published by the Deutsche Nationalbibliothek: The Deutsche Nationalbibliothek lists this publication in the Deutsche Nationalbibliografie; detailed bibliographic data are available in the Internet at http://dnb.d-nb.de.

Any brand names and product names mentioned in this book are subject to trademark, brand or patent protection and are trademarks or registered trademarks of their respective holders. The use of brand names, product names, common names, trade names, product descriptions etc. even without a particular marking in this works is in no way to be construed to mean that such names may be regarded as unrestricted in respect of trademark and brand protection legislation and could thus be used by anyone.

Publisher: Südwestdeutscher Verlag für Hochschulschriften Aktiengesellschaft & Co. KG
Dudweiler Landstr. 99, 66123 Saarbrücken, Germany
Phone +49 681 37 20 271-1, Fax +49 681 37 20 271-0
Email: info@svh-verlag.de

Printed in the U.S.A.
Printed in the U.K. by (see last page)
ISBN: 978-3-8381-1646-4

Copyright © 2010 by the author and Südwestdeutscher Verlag für Hochschulschriften Aktiengesellschaft & Co. KG and licensors
All rights reserved. Saarbrücken 2010

CONTENTS

1. **General introduction and perspectives: Effects of human disturbance on tropical epiphyte communities**   1
   Tropical montane forests and vascular epiphytes   2
   Vascular epiphytes and human disturbance: state of knowledge   3
   General objectives and outline of chapters   5
   Perspectives   6

2. **Diversity of vascular epiphytes on isolated remnant trees in the montane forest belt of southern Ecuador**   9
   Abstract   10
   Introduction   11
   Study area   12
   Methods   13
   Results   16
   Discussion   19

3. **Diversity of dry forest epiphytes across a gradient of human disturbance in the tropical Andes**   24
   Abstract   25
   Introduction   26
   Methods   27
   Results   33
   Discussion   35

4. **Spatial distribution and abundance of epiphytes across a gradient of human disturbance in an Interandean dry valley, Ecuador**   40
   Abstract   41
   Introduction   42
   Methods   44
   Results   47
   Discussion   50

5. **Is the resilience of epiphyte assemblages to disturbance a function of local climate?** — **57**
   - Abstract — 58
   - Introduction — 59
   - Methods — 61
   - Results — 63
   - Discussion — 65

6. **Seedling establishment of vascular epiphytes on isolated and enclosed forest trees in an Andean landscape, Ecuador** — **70**
   - Abstract — 71
   - Introduction — 72
   - Methods — 73
   - Results — 78
   - Discussion — 79

7. **Increased mortality of vascular epiphytes on isolated trees following forest clearance in moist montane South Ecuador** — **85**
   - Abstract — 86
   - Introduction — 87
   - Methods — 88
   - Results — 92
   - Discussion — 95

8. **References** — **104**

9. **Abstracts** — **128**
   - Abstract — 128
   - Resumen — 131
   - Zusammenfassung — 135

10. **Appendices** — **140**

11. **Acknowledgments** — **154**

Chapter 1

GENERAL INTRODUCTION AND PERSPECTIVES: EFFECTS OF DISTURBANCE ON TROPICAL EPIPHYTE COMMUNITIES

TROPICAL MONTANE FORESTS AND VASCULAR EPIPHYTES

Cloud forests have commonly been defined as forests that receive substantial moisture input from fog at a regular basis (Bruijnzeel 2005). Typically, cloud forests cover the slopes of tropical mountains along elevational bands corresponding to the layers of cloud formation (but see Gradstein 2006) resulting from the convective uplift of air masses (Shuttleworth 1977; Rollenbeck et al. 2005). In such tropical montane cloud forests, moisture tends to be plentiful throughout the year, and epiphytic vascular plants and bryophytes attain high diversity, their greatest abundance and, presumably, functional importance. These forests represent the 'classical' tropical montane forest, and for many authors the term 'cloud forest' has become almost synonymous of tropical montane forests as a whole. The lush epiphytic vegetation of cloud forests has been addressed in numerous studies (Hofstede et al. 1993; Nadkarni 1984, Höft & Höft 1993; Engwald 1999; Rudolph et al. 1998; Freiberg & Freiberg 2000; Nowicki 2001; Webster & Rhode 2001). However, the complex topography of tropical mountains fosters a broad variety of climates and vegetation types, including many forest types that do not receive much fog. Where 'horizontal precipitation' (Vogelmann 1973) from cloud-combing cannot compensate for lack of rainfall, epiphytes face more frequent and stronger droughts. The epiphytic vegetation of such montane forests tends to be less lush and prominent, and has received less attention in epiphyte research, even though these forests can be equally diverse – even mid-sized trees can foster exceptionally high species numbers of vascular epiphytes (Werner et al. 2005; chapter 2). Such a montane forest in southern Ecuador was chosen as one of the two sites for this study, and is treated in chapters 2, 6 and 7.

With increasing aridity, epiphyte diversity eventually drops markedly (Gentry & Dodson 1987a, 1987b; Kreft et al. 2004). Even though distinctly seasonal montane forests have relatively low epiphyte diversity, they can foster high levels of endemism, and epiphytes may continue to play a major role in ecosystem functioning (chapter 4). Montane dry forests occur throughout the Tropics and Subtropics, and the challenges of prolonged drought have promoted a number of striking adaptations among epiphytes (e.g., CAM metabolism, deciduousness, leaf trichomes for water absorbance). Since little is known about this extreme environment as a habitat for epiphytes, such a dry forest was chosen as the second site for this study (chapters 3 and 4).

Chapter 1 — General introduction and perspectives

VASCULAR EPIPHYTES AND HUMAN DISTURBANCE: STATE OF KNOWLEDGE

The keen and extensive observations of Schimper (1888) mark the starting point of tropical epiphyte research. But only the introduction of mountaineering techniques to ecological sciences (Denison 1973; Perry 1978) ended the era of opportunistic scavenging of fallen trees and opened the path for quantitative ecological studies. Johansson (1974) provided a first extensive quantitative study on the ecology of vascular epiphytes communities. His classical work documented pronounced successional dynamics in epiphyte communities that inspired intensified research efforts (Madison 1979; Sudgen & Robbins 1979; Yeaton & Gladstone 1982; Catling & Lefkovitch 1989; ter Steege & Cornelissen 1989). It quickly became apparent that epiphytes contribute substantially to tropical plant diversity at global and regional, but particularly at local scale, where up to 50% of vascular plant species may be epiphytes (Madison 1977; Kelly 1985; Kress 1986; Homeier & Werner, in press; Lehnert et al., in press).

The revelation that epiphytes constitute a major fraction of global plant diversity fuelled considerable interest in the latitudinal and altitudinal distribution of epiphyte diversity, which has cumulated in a good understanding of global patterns of alpha diversity (e.g., Kreft et al. 2004). Horizontal and vertical gradients of humidity have also been studied extensively with respect to epiphyte diversity and have corroborated the paramount importance of moisture availability as a predictor of epiphyte diversity and floristic composition (e.g., Johansson 1974; Sudgen & Robbins 1979; Gentry & Dodson 1987a, 1987b). Another newly tackled field was the functional role of epiphytes, culminating in a series of studies on their biomass (e.g., Tanner 1980; Nadkarni 1984; Veneklaas et al. 1990; Ingram & Nadkarni 1993; Freiberg & Freiberg 2000), water storage capacity (e.g., Pócs 1980; Kürschner & Parolly 2004), mineral contents (Hofstede et al. 1993; Nadkarni 1986) and biotic interactions (Nadkarni & Matelson 1989; Vance & Nadkarni 1990). These excitingly new aspects absorbed considerable interest and it took another 20 years before human disturbance effects began to move into the scope of epiphyte researchers.

At present, some 30 years after the birth of canopy research, our understanding of how tropical epiphytes respond to human disturbance remains fragmentary. The literature on cool-temperate and boreal epiphytes (lichens, bryophytes) is already ample and in rapid growth. Unfortunately, there is still little evidence that results gained from such studies are applicable to vascular epiphytes, which are essentially (and in striking contrast to the great majority of lichens and bryophytes) desiccation-intolerant.

Our understanding of the mechanisms and processes governing tropical vascular epiphyte communities is essentially based on descriptive studies that rely on time-for-space replacement (Laube & Zotz 2006). Long-term studies have been used widely for terrestrial plant communities, providing unique insights into their dynamics, but are virtually nonexistent for epiphytes (Zotz 2004a; Laube & Zotz 2006). Experimental approaches are similarly scarce. For instance, besides few studies on hemiepiphytic figs, experimental studies on early establishment have virtually exclusively dealt with tillandsioid bromeliads. The bulk of bromeliads prefer early-successional substrates, high light levels, and are conspicuously hardy and resilient to disturbance, related to a unique set of adaptations (e.g., foliar trichomes, cisterns; Benzing 1990; Barhlott et al. 2001; Werner et al. 2005; Hietz et al. 2006). Despite of being one of the largest and most important epiphyte groups in the Neotropics, they can therefore not be considered representative for epiphytic communities.

The limited number of case studies on tropical epiphyte assemblages has yielded a wide array of responses to aspects of human disturbance, ranging from unchanged to greatly impoverished (see chapter 5). Within single taxa, responses can even vary more greatly. For example, in Andean secondary moist forest bromeliads can exhibit significantly higher (Barthlott et al. 2001) or lower (Krömer & Gradstein 2003) diversity than in adjacent primary forest.

Indirect evidence suggests a rather limited number of major drivers (e.g., microclimate, dispersal constraints, substrate properties, grade of disturbance, and the time-span of subsequent recovery). However, field studies usually have to deal with the response of diverse communities to a complex and unique mixture of these parameters. Most of these factors are interrelated and difficult to quantify, and have never been addresses thoroughly in studies on epiphyte assemblages of non-primary habitats.

The present study focussed primarily on scattered trees isolated in anthropogenic land use matrices. Such trees constitute keystone structures that offer refuge and enhance connectivity for forest organisms, and provide nuclei of regeneration (Janzen 1988; Hietz 2005; Wolf 2005; Manning et al. 2006; Zahawi & Augspurger 2006). Moreover, isolated trees lend themselves as an excellent model system for the studying of human disturbance effects. Isolated trees constitute the smallest possible forest fragment (cf. Williams-Linera et al. 1995) exposed to multiple (maximum) physical edge effects. They are of well-defined and comparable size, easily replicable, and subjected to quantifiable dispersal limitations. Unlike degraded or secondary forests, isolated trees neither differ between sites in the relative degree of their physical exposure, nor do they differ from undisturbed forest regarding relevant host parameters such as age, surface area or bark characteristics, unless biased in size or taxonomic composition. These virtues were first recognized by Richter (1991, 2003), who used epiphytes on isolated trees as indicators of local climate for the elaboration of regional climate maps in the Neotropics.

GENERAL OBJECTIVES AND OUTLINE OF CHAPTERS

The present dissertation aims at increasing our knowledge regarding the responses of epiphyte communities to different aspects of human disturbance. Chapters are largely prepared and arranged as manuscripts for submission to journals or have already been submitted. However, some of the discussion sections are relatively extensive while methods sections have been shortened in order to avoid repetition; moreover, the editing may vary in detail between chapters. The specific objectives of this study are addressed in the following paragraph.

Chapter 2 provides first results of a descriptive study on the development of vascular epiphyte assemblages on remnant trees 10–30 yr following their isolation in Andean pastures on the moist eastern Andean flank. Analysed are abundance, diversity and floristic composition on isolated trees vs. forest trees, and the distribution of these parameters along the trunk-branch-twig trajectory.

Chapter 3 describes how different types and grades of disturbance affect diversity and floristic composition of epiphytic vascular plants and bryophytes in a tropical montane dry forest landscape. This study allows novel insights in that it constitutes the first attempt to quantify disturbance effects on dry forest epiphytes, the first direct comparison of disturbance effects between epiphytic bryophytes and vascular plants, and the first gradient analysis of disturbance effects on epiphytes. Chapter 4 provides supplementary data on spatial distribution and abundance relations of epiphytes obtained from the same study, as well as some general observations on the natural history of the study area.

In chapter 5, a hypothesis is tested which aims to unite and explain hitherto contradictive results yielded by different studies on the consequences of human disturbance on vascular epiphytes. This hypothesis is based on the assumption that microclimatic changes are a principal driver of epiphyte diversity in disturbed habitats, and predicts that the (relative) magnitude of impact of structural forest disturbance on epiphyte diversity is a non-linear function of local climate.

Chapters 6 and 7 present the results of field trials in southeast Ecuador aimed at sheding light on the mechanisms behind the pronounced long-term impoverishment of vascular epiphyte assemblages on isolated remnant trees that are described in chapter 2. Specifically, I wanted to test the hypotheses that mid-term impoverishment is primarily caused by altered microclimate and that physical 'edge' effects affect both recruitment (chapter 5) and the persistence of well-established plants (chapter 6).

PERSPECTIVES

The results of the descriptive studies presented in chapters 2 to 4 strongly suggest an overarching influence of (micro-)climatic constraints on the development of tropical epiphyte communities in anthropogenic landscapes. From a short- and mid-term perspective this conclusion is lent strong support by direct evidence from experimental data on well-established plants (chapter 7). These findings are in contrast to a number of studies that concluded paramount influence of dispersal limitations for the maintenance of epiphyte diversity in anthropogenic landscapes (e.g., Peck & McCune 1997; Sillett et al. 2000; Zartman 2003; Öckinger et al. 2005; Wolf 2005; Cascante 2006; Zartman & Nascimeto 2006). Recent studies (Pharo & Zartman 2007; Werth et al. 2007) have emphasised the difficulty to separate regional (dispersal constraints) from local factors (microclimate, substrate quality) in field studies, and suggest that many studies may have overestimated the former at the cost of the latter. These difficulties are mirrored in chapter 6. Although the field trial described in this chapter could show that the establishment on isolated trees was greatly reduced and compositionally biased relative to closed forest, it failed to disentangle the roles of regional and local factors. The outcomes strongly suggest additive effects of dispersal constraints and elevated early seedling mortality due to a changed microclimate, but the weights of these factors as predictors of epiphyte abundance, diversity and composition remained unclear. The fact that mortality of well-established plant individuals on freshly isolated trees was drastically increased (chapter 7) raises the question as of which aspects of increased physical exposure act primarily upon plant performance. 'Physical edge effects' can conveniently be split into two principal components, namely solar radiation and wind turbulence. These two factors may have substantial detrimental effects on plan performance directly (mechanical injury, photoinhibition, UV-damage). However, light levels also affect temperatures of air, substrate and plant tissue, which potentially affects carbon balance, physiological processes as well as the integrity of plant tissues. In combination with wind turbulence, light also strongly influences water availability, which appears to be a bottleneck for many epiphytes (Zotz & Tyree 1996, Zotz & Hietz 2001). To further complicate matters, edge effects also affect substrate quality and, probably, mycorrhizal fungi, which provide many vascular epiphytes with water and nutrients (Fig. 1).

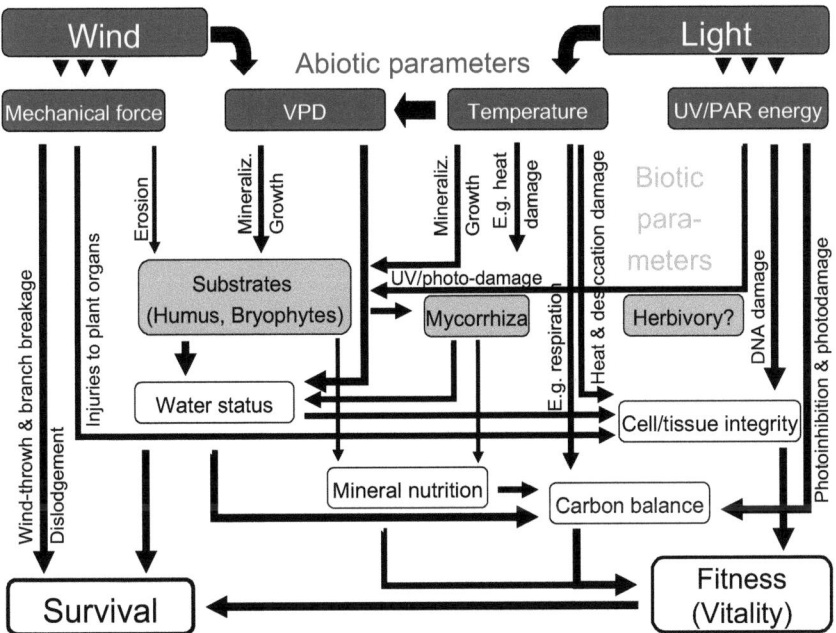

Figure 1. Schematic overview of the pathways by which physical edge effects (s.l.) may affect vascular epiphyte performance following canopy disrupture.

Understanding how these complexly interrelated and cascading effects affect epiphyte performance is a major challenge. Much can be learned from studies of human disturbance effects on the response of epiphytes to climate change. Just as in disturbed habitats, temperatures and vapour pressure deficit are projected to rise significantly over vast areas with climate change, but levels of light and wind are not. How effective are climate change-relevant factors on the short term? How do they affect survival and fecundity of epiphytes? To which extent and by which means (e.g., pigmentation, anatomy, physiology) can plants adjust over time? How exactly and to which degree may early seedling requirements create a bottleneck in the life-cycle of epiphytes in a changing environment (compare discussion in chapter 6)? Additional field studies are needed to arrive at a more generalized synthesis, extending beyond local scenarios (see chapter 5). In combination with ecological and ecophysiological experiments, these studies may lead to a more thorough understanding of the processes governing epiphyte communities in non-natural habitats.

Chapter 2

DIVERSITY OF VASCULAR EPIPHYTES ON ISOLATED REMNANT TREES IN THE MONTANE FOREST BELT OF SOUTHERN ECUADOR

WITH J. HOMEIER & S. R. GRADSTEIN

ABSTRACT

We studied the diversity of vascular epiphytes on isolated remnant trees of pastures in southern Ecuador. The objective of this study was to document the importance of remnant trees for the survival of vascular epiphytes following forest clearance. Twenty-one canopy trees (15 remnant trees, 6 forest trees) were divided into five zones following Johansson (1974) and climbed with the single rope technique. Recorded parameters include floristic composition, diversity, abundance, and spatial distribution of epiphytes. Bromeliaceae, Orchidaceae, Piperaceae, Polypodiaceae were relatively well represented on remnant trees in terms of species richness and abundance, whereas other families such as Dryopteridaceae, Ericaceae, Grammitidaceae and Hymenophyllaceae were poorly represented or absent. Abundance and diversity of epiphytes were significantly lower on remnant trees than on forest trees. Impoverishment was greatest on the stem base and in the outer crown, and least in the inner crown of the host trees. We postulate that microclimatic changes and increased drought stress following the isolation of the remnant trees resulted in lowered rates of establishment and survival of vascular epiphytes.

Key words: deforestation, diversity, Ecuador, epiphytes, remnant trees, spatial distribution, tropical montane forest, tropical pastures

INTRODUCTION

Most studies on tropical forest fragmentation focus on forest fragments only, neglecting the characteristics of the surrounding habitats (Saunders et al. 1991). Consequently, it remains largely unknown how tropical forest organisms respond to habitat characteristics outside remnant fragments (Guevara 1995). Meanwhile, it has become increasingly apparent that understanding how species are affected by fragmentation requires information on their responses to all landscape components, including the forest-intervening matrices (Gascon et al. 1999).

This paper deals with vascular epiphyte assemblages on isolated remnant trees (IRTs) occurring in tropical pastures. Vascular epiphytes abound in tropical forests, especially in montane ones, and are often highly sensitive to anthropogenic disturbance (King & Chapman 1983; Hickey 1994; Turner et al. 1994; Barthlott et al. 2001; Krömer 2003; Krömer & Gradstein 2003). However, knowledge of epiphyte assemblages on IRTs remains very poor.

Epiphytic vegetation provides important resources and habitat for a wealth of animals and micro-organisms (Vance & Nadkarni 1990; Paoletti et al. 1991; Greeney 2001; Stuntz et al. 2002). These include insectivorous, nectarivorous and frugivorous vertebrates, especially bats and birds (Nadkarni & Matelson 1989; Castañeda 2001; Fleming et al. 2005). The latter play a key role in gene flow between forest fragments and regenerating forest patches on abandoned pastures and fields (McDonnell & Stiles 1983; Guevara et al. 1992). Generally, birds and bats are reluctant to enter or cross open landscapes unless these areas offer significant reward (Charles-Dominique 1986; Nepstad et al. 1990; Githiru et al. 2002). Given suitable attractiveness, IRTs can stimulate movement of birds and bats across the forest border and function as catalysts for forest regeneration (Guevara et al. 1986; Janzen 1988; Cardoso da Silva et al. 1996; Duncan & Chapman 1999; Carrière et al. 2002a, 2002b).

The objective of this study was to describe the diversity and abundance of vascular epiphytes on IRTs. Specifically, we wanted to (1) document the effects of isolation on species composition, (2) determine possible causes of the structure and diversity of the vascular epiphyte assemblage on remnant trees, and (3) document the importance of remnant trees for the survival of vascular epiphytes following forest clearance.

STUDY AREA

The study was carried out in the valley of the Río San Francisco, southern Ecuador (3° 58' S, 79° 04' W), near the Estación Científica San Francisco (ECSF) at ca. 1800-2200 m elevation. The study area is situated within the Cordillera El Consuelo, forming part of the eastern range of the Ecuadorian Andes and bordering Podocarpus National Park. The region has been identified as a center of endemism and diversity for major groups of organisms such as birds, vascular plants or bryophytes (Fjeldså 1995; Borchsenius 1997; Navarrete 2000; Valencia et al. 2000; Parolly et al. 2004).

The relief is highly structured by deeply incised ravines, steep slopes of 20-55° inclination, and narrow ridge-tops. Landslides are very common and result in a complex mosaic of successional stages of vegetation. Soils are very heterogeneous but are generally shallow, highly acidic and very poor in basic cations and effective cation exchange capacity (Schrumpf et al. 2001).

At 1950 m a.s.l. mean temperature is 15.5°C and average air humidity is 86%. Annual precipitation averages slightly above 2000 mm (Emck 2007). Rainfall seasonality is not very pronounced; differences between years exceed those within years (R. Rollenbeck, pers. comm.). April – June are generally the wettest months while September – February tend to be drier. Since the beginning of climate recording in 1998, periods without precipitation longer than one week have been recorded only during November – January. The San Francisco valley experiences slight lee- and foehn-effects (P. Emck, pers. comm.). Fog is uncommon throughout the year (F. Werner, pers. obs.).

Primary forests on the north-facing slopes are generally of low stature, with canopy height exceeding 15-20 m only in ravines. Physiognomic differences between ridges, slopes and ravines are large (Homeier et al. 2002). Forests on the south-facing slopes were largely converted to cattle pastures ca. 12-30 yrs prior to sampling, with loosely-spaced occurrence of isolated remnant trees (IRTs). However, two of the sampled IRTs were isolated as recently as 2 and 5 years prior to sampling (Appendix 1). *Cedrela montana* and *Tabebuia chrysantha* are the main remnant tree species, being preserved, at least temporarily, for their valuable timber. Trees surviving slash-and-burn clearance generally exhibit healthy growth. Forest regeneration is prevented by burning of pastures during dry periods (Hartig & Beck 2002). Remnant vegetation or secondary forest occurs in scattered patches, mostly in narrow bands along ravines.

METHODS

Fifteen IRTs in pasture on the north-facing slope of the San Francisco valley and 6 canopy trees at similar elevation in nearby primary forest on the south-facing slope were sampled.

Distances between IRTs and intact forest varied from approximately 100-500 m. Trees were selected randomly among accessible canopy trees of 30-50 cm diameter at breast height (DBH); forest trees 4-6 were sought for to avoid bias by host identity. Both sub-samples have similar shares of trees from ridges, slopes and ravines.

Access to tree crowns was achieved using the single rope technique (Perry 1978). In a few cases specimens were gathered by employing a hooked pole or by cutting off minor branches. Voucher specimens were deposited in AAU, ECSF, MO, SEL, QCA and QCNE.

Tree height, DBH, location, elevation, and time elapsed since isolation (age of clearing) were recorded. The latter was determined by interviewing land users. Epiphytes were sampled in each of 5 vertical tree zones following a zonation scheme slightly modified after Johansson (1974). Johansson-zone 1 (JZ 1) stretches from 0.25 m up to 3.0 m, JZ 2 from 3 m above ground to the first major ramification, JZ 3 comprises major branches > ca.12 cm in diameter (inner crown), JZ 4 branches 12-5 cm in diameter (middle crown), and JZ 5 branches < 5 cm in diameter (outer crown). Surface areas of zones 3 - 5 are about equal.

Vascular epiphytes sampled included facultative and obligate holoepiphytes (sensu Benzing 1990), primary and secondary hemiepiphytes (Todzia 1986; Putz & Holbrook 1989), and accidental epiphytes (Benzing 1990). Non-hemiepiphytic climbers, hemiparasites and seedlings were excluded. Because of the common occurrence of clumped species, "stands" instead of individuals were recorded (stand = group of stems or plants spatially separated from another group of the same species by an area on the tree devoid of epiphytes or occupied by another species; Sanford 1968). Covers of bryophytes, lichens and substrate accumulations > 1 cm thick were estimated in steps of 5% in relation to tree surface. Substrate accumulations consisted of dead organic matter in various stages of decomposition ("crown humus"; Jenik 1973) and living bryophytes and lichens.

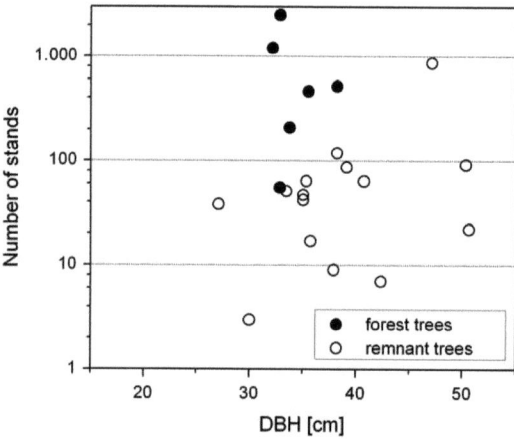

Figure 1. Abundance of epiphytes in relation to tree size (DBH).

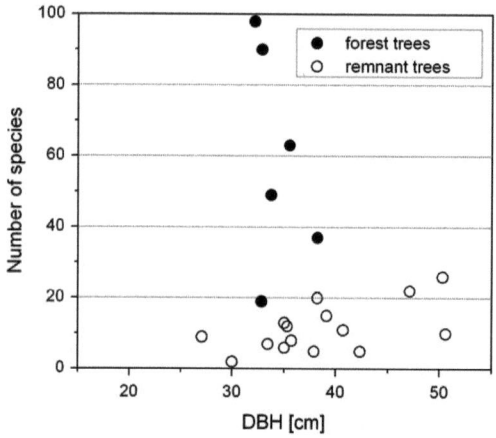

Figure 2. The number of epiphyte species in relation to tree size (DBH).

Statistical analysis was performed by two-tailed Mann-Whitney U-test and the Spearman rank-correlation test without transformations. Sørensen similarities between forest trees and IRTs were calculated for whole trees and Johansson-zones. Nonmetric multidimensional scaling (NMDS) was applied to the resulting matrices of similarity, here displayed as two-dimensional scatterplots.

Table 1. Floristic composition of epiphytes from the 6 forest trees (FTs) and 15 IRTs ordered by families.

	Richness [no. species]					Abundance [no. stands]				
	Total			Relative [%]		Total			Relative [%]	
	FTs	IRTs	Sum	FTs	IRTs	FTs	IRTs	Sum	FTs	IRTs
Alzateaceae	1	–	1	<1	–	1	–	1	<1	–
Araceae	6	2	8	3	3	7	4	11	<1	<1
Araliaceae	2	–	2		–	2	–	2	<1	–
Asclepiadaceae	1	–	1	<1	–	1	–	1	<1	–
Asteraceae	2	–	2	<1	–	4	–	4	<1	–
Bombacaceae	1	–	1	<1	–	1	–	1	<1	–
Bromeliaceae	23	13	25	10	19	502	374	876	10	24
Cactaceae	1	–	1	<1	–	1	–	1	<1	–
Clusiaceae	3	–	3	1	–	4	–	4	<1	–
Cunoniaceae	1	–	1	<1	–	2	–	2	<1	–
Cyclanthaceae	1	–	1	<1	–	5	–	5	<1	–
Dryopteridaceae	9	1	9	<1	1	190	1	191	4	<1
Ericaceae	13	1	13	6	1	84	1	85	2	<1
Gesneriaceae	1	–	1	<1	–	1	–	1	<1	–
Grammitidaceae	13	1	13	<6	1	565	2	567	11	<1
Hydrangeaceae	1	–	1	<1	–	2	–	2	<1	–
Hymenophyllaceae	10	–	10	4	–	74	–	74	1	–
Lentibulariaceae	1	–	1	<1	–	592	–	592	12	–
Marcgraviaceae	2	–	2	<1	–	3	–	3	<1	–
Melastomataceae	5	–	5	2	–	7	–	7	<1	–
Moraceae	1	2	3	<1	3	1	3	4	<1	<1
Orchidaceae	105	31	120	47	46	2802	1004	3806	56	65
Piperaceae	10	7	12	4	10	52	24	76	1	2
Polypodiaceae	9	7	12	4	10	42	115	157	<1	8
Rubiaceae	1	–	1	<1	–	1	–	1	<1	–
Solanaceae	–	2	2	–	3	–	6	6	–	<1
Urticaceae	1	–	1	<1	–	2	–	2	<1	–
Vittariaceae	1	–	1	<1	–	26	–	26	<1	–
Total	225	67	253			4974	1534	6508		

RESULTS

Composition and diversity

A total of 6508 stands representing 253 species of vascular epiphytes (86 genera, 28 families) was recorded. The 6 sampled forest trees harbored 4974 stands of 225 species (80 genera, 27 families), the 15 sampled IRTs 1534 stands of 67 species (30 genera, 10 families). Bromeliaceae, Orchidaceae, Piperaceae and Polypodiaceae were best represented on remnant trees regarding species richness and abundance (Table 1). Abundance of Orchidaceae on IRTs was largely due to the succulent *Dryadella werneri*, constituting 73% of all orchids. Compared with forest trees, species richness on IRTs was most strongly reduced in Dryopteridaceae (*Elaphoglossum*) (89%), Ericaceae (92%), Grammitidaceae (92%) and Hymenophyllaceae (100%), least in Bromeliaceae (44%), Piperaceae (30%) and Polypodiaceae (22%). Species with considerable abundance on IRTs included *Tillandsia complanata* and *Tillandsia. tovariensis* (Bromeliaceae), *Dryadella werneri, Epidendrum stangeatum, Epidendrum* cf. *zosterifolium* and *Prosthechea grammatoglossa* (Orchidaceae), and *Pleopeltis macrocarpa* and *Polypodium remotum* (Polypodiaceae) (see Appendix 2).

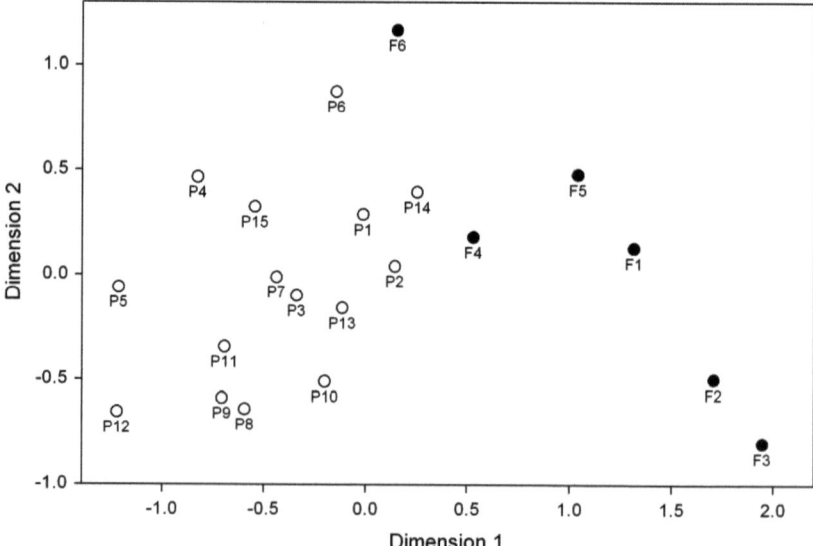

Figure 3. Nonmetric multidimensional scaling plot (first two dimensions) of epiphyte similarity based on Sørensen index for entire host trees. Closed circles (F1-6) represent forest trees, open circles (P1-15) IRTs.

The number of epiphyte stands on single trees varied greatly (Fig. 1). Forest trees held 55-2519 stands (mean 828.7; median 490.0), IRTs 3-872 stands (mean 102.3; median 47.0). Total abundance and species density of epiphytes were significantly lower on IRTs, both on whole trees and in Johansson-zones, compared with forest trees (Table 2).

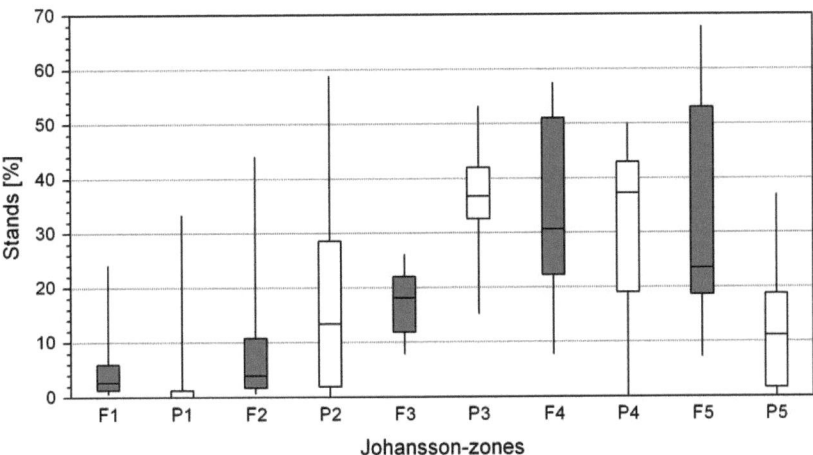

Figure 4. Relative abundance (% of the hosts' stand numbers) recorded along the Johansson-zones: forest trees (grey) vs. IRTs (white).

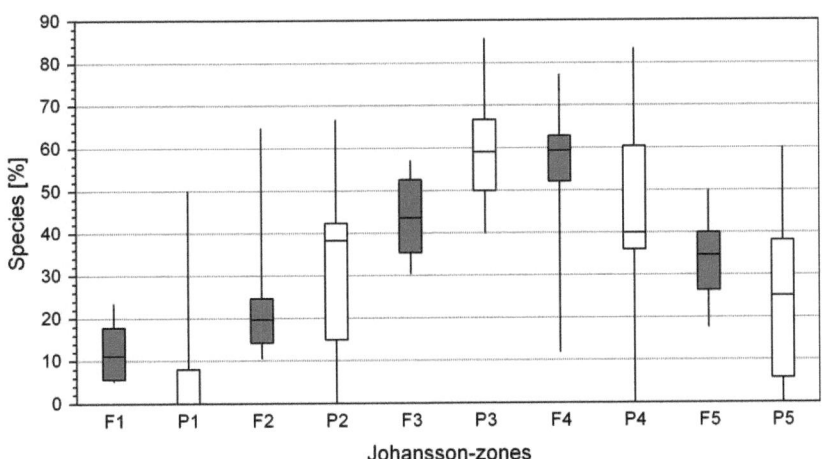

Figure 5. Relative species density (% of the hosts' total species richness) recorded along the Johansson-zones on forest trees (grey) vs. IRTs (white).

IRTs harbored 2-26 species (mean 10.5; median 10.0), forest trees 19-98 (mean 59.3; median 56.0) (Fig. 2). Diversity (Shannon, Simpson) was significantly lower on IRTs ($p < 0.001$, $n = 21$ and $p < 0.05$, $n = 21$ respectively; U-test). Species richness and abundance were correlated positively with covers of bryophytes and substrate accumulations, and negatively so with lichen cover (Table 3). Epiphyte assemblages on forest trees and IRTs were grouped as separate floristic units by NMDS (Fig. 3). The first dimension well reflects species richness with the poorest hosts (trees P4, P5 and P12) on the left and the richest trees (F1, F2 and F3) on the right of the graph.

Table 2. Species richness and abundance (absolute and relative respectively) along Johansson-zones. Forest trees vs. IRTs. Significance levels of Mann-Whitney U-test. All $n = 20$ except for JZ 2 (20).

	JZ 1	JZ 2	JZ 3	JZ 4	JZ 5	Totals
Species richness	< 0.001	< 0.01	< 0.01	< 0.01	< 0.001	< 0.01
Relative richness	< 0.05	n.s.	< 0.05	n.s.	n.s.	–
Abundance	< 0.001	< 0.01	< 0.05	< 0.05	< 0.01	< 0.01
Relative abundance	< 0.05	n.s.	< 0.05	n.s.	< 0.05	–

Spatial distribution

Mean relative abundance was highest in JZ 4 and 5 on forest trees and in JZ 3 and 4 on IRTs (Fig. 4). Relative abundance and species density on IRTs compared with forest trees were significantly lower in JZ 1 but higher in JZ 3 (Table 2). In addition, relative abundance on IRTs was significantly lower in JZ 5. Mean relative species density on forest trees was highest in JZ 4, on IRTs in JZ 3 (Fig. 5). NMDS of the Johansson-zones based on assemblage structure clearly separated the two habitats (Fig. 6). Within each habitat, epiphyte assemblages of crown-zones (JZ 3-5) were grouped closely together, those of lower stems (JZ 1) were well isolated. Upper stems (JZ 2) of forest trees, finally, were clearly distinct from crown-zones, whereas this zone showed great similarity to the crown in remnant trees.

DISCUSSION

Composition and diversity

Floristic composition of vascular epiphytes in the investigated forest shows close resemblance to other moist neotropical mid-elevation forests (e.g., Ibisch 1996; Ingram et al. 1996; Engwald 1999; Freiberg & Freiberg 2000; Krömer 2003; Krömer & Gradstein 2003) and species richness is very high (see also Bussmann 2001). In comparison, the epiphytic flora on IRTs in the study area is impoverished and monotonous. Bromeliaceae, Orchidaceae, Piperaceae, and Polypodiaceae, all being rich in drought-tolerant species, were relatively species-rich and abundant, whereas Ericaceae, Dryopteridaceae, Grammitidaceae and Hymenophyllaceae, all being common forest elements, were scarce or lacking on IRTs (Table 1). These findings agree with recent studies in Bolivia (Ibisch 1996; Krömer & Gradstein 2003). In a strongly seasonal montane forest in Bolivia (6-8 arid months), the epiphytic flora consisted of Bromeliaceae, Cactaceae, Orchidaceae, Piperaceae, and Polypodiaceae (Ibisch 1996). Krömer & Gradstein (2003) found that Piperaceae and Polypodiaceae were well represented in open fallows in moist submontane Bolivia (1500-2000 mm/an. precipitation; 2-3 arid months), while Bromeliaceae and Orchidaceae exhibited considerably reduced species richness compared with the primary forest. Dryopteridaceae (*Elaphoglossum*), Grammitidaceae, and Hymenophyllaceae were virtually lacking in the fallows.

Table 3. Spearman rank-correlations between selected parameters ($n = 21$). None of the given parameters correlates with altitude, DBH, tree height or elapsed time since isolation.

	Species richness	Abundance	Shannon H'	Simpson D
Species richness	–	0.936 **	0.765 **	- 0.412
Abundance	0.936 **	–	0.567 **	- 0.237
Shannon H'	0.785 **	0.860 **	–	- 0.823 **
Simpson D	0.692 **	0.682 **	- 0.823 **	–
Bryophyte cover	0.605 **	0.651 **	0.270	- 0.015
Lichen cover	- 0.534 *	- 0.545 *	- 0.211	0.047
Substrate accumulation cover	0.692 **	0.682 **	0.488 *	- 0.260

*significant at $p < 0.05$; ** significant at $p < 0.01$.

Reduction of species diversity on IRTs was paralleled by reduced bryophyte cover, but correlated negatively with lichen cover, which was increased on IRTs. Decreased diversity and cover of epiphytic bryophytes in IRT crowns is related to increased evaporation and insulation, as has previously been documented by Sillett et al. (1995). Bryophyte cover tends to increase with humidity (Gradstein & Pócs 1989) while lichens avoid excessive humidity and shading (Sipman & Harris 1989; Gradstein 1992). Thus the observed patterns strongly suggest increased drought stress as the principal agent for the compositional shifts and general impoverishment in terms of species richness and abundance of assemblages on IRTs.

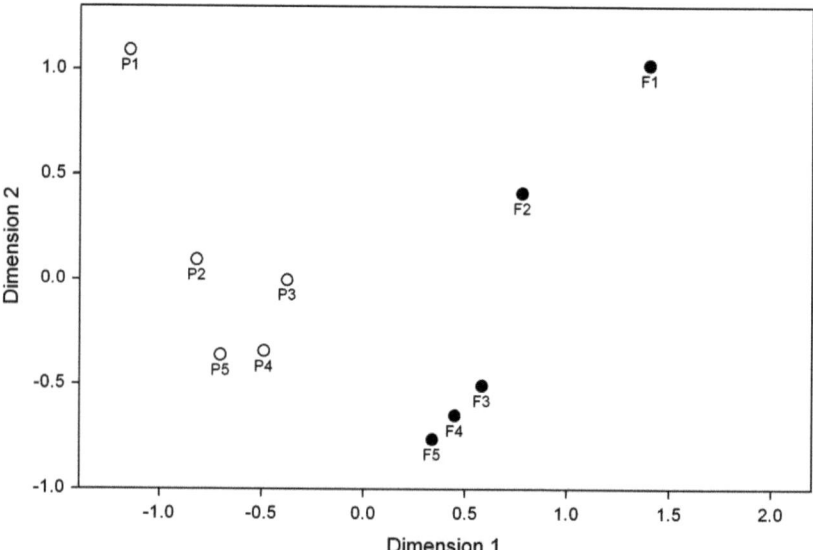

Figure 6. Nonmetric multidimensional scaling plot (first two dimensions) of epiphyte similarity based on Sørensen index for the five Johansson-zones. Closed circles (F1-5) represent the pooled respective Johansson-zones of forest trees, open circles (P1-5) those of IRTs.

Numerous workers have noted the importance of humidity to epiphyte diversity (e.g., Gentry 1988; Ek et al. 1997; Kreft et al. 2004). Their extreme sensitivity to drought as a consequence of an aerial life style makes epiphytes important as prime indicators for mesoclimates and climate change (Richter 1991; Lugo & Scatena 1992; Nadkarni 1992; Benzing 1998; Richter 2003). One of the most striking patterns shown by epiphytes is the large decrease in both numbers of species and individuals in drier habitats (Gentry & Dodson 1987a). When transplanted to warmer and drier conditions, epiphytes responded with higher leaf mortality, lower leaf production and reduced longevity (Nadkarni & Solano 2002).

Flores-Palacios & García-Franco (2004) reported similar impoverishment of epiphyte assemblages on IRTs in montane Mexico. The site is moderately moist and experiences a distinct dry season (1650 mm/an. precipitation with 7 dry months). Decreased diversity on IRTs is not a general trend though. In moist areas of lowland southern Mexico and lower montane northern Ecuador, species richness on IRTs was similar to that on forest trees (Larrea 1995, 1997; Hietz-Seifert et al. 1996), although floristic composition was more uniform on IRTs than on forest trees in at least one of these studies (Larrea 1997). Both sites show high precipitation and little seasonality with > 4000 mm/an. precipitation and the two driest months with ca. 100 mm, and ca. 3500 mm/an. and no arid months respectively (P. Hietz and H. Greeney, pers. comm.). We suggest that in aseasonal, perhumid climates impoverishment of epiphyte vegetation on IRTs compared with nearby forest is less severe than in moderately seasonal climates such as the present study area. In areas with perhumid climates, high air humidity may be maintained in open habitats following deforestation, allowing for high epiphyte species richness on IRTs, even though considerable turnover may follow isolation. Under slightly seasonal conditions with moderate levels of drought stress, however, species richness of epiphytes is high only in the forest, where high air humidity is maintained under the closed canopy. Opening up of the canopy in these areas leads to significant changes in the air humidity regime (F. Werner, unpubl. data) and subsequent impoverishment in abundance and diversity of epiphytes. Interestingly, in arid regions where total annual precipitation is low and epiphyte diversity limited, epiphyte assemblages on IRTs are often relatively unchanged compared with the forest (F. Werner, pers. obs.). It thus appears that loss of diversity on IRTs in tropical regions is most severe in areas with a moderately seasonal climate. However, fog appears disproportionately beneficial for epiphytes on IRTs, complication the interpretation of precipitation effects wherever it occurs regularly.

Vertical distribution

Vertical stratification of epiphytes on forest trees in relation to changes in microclimatic conditions along the tree has often been described (e.g., Johansson 1974; Sudgen & Robbins 1979; Kelly 1985; ter Steege & Cornelissen 1989). Data on remnant trees are very scarce, however. On IRTs in Mexico branches with diameters less than 5 cm were only sparsely colonized by vascular epiphytes (Hietz-Seifert et al. 1996). A similar pattern was found in this study (Fig. 4; Table 2). Inner crowns of IRTs in the study area, however, were significantly richer in terms of relative species richness and abundance than inner crowns of forest trees). The uneven distribution of species diversity in IRT crowns may reflect reduced rates of successful colonization after isolation.

Nadkarni (1992) reported paucity of epiphytes on trunk bases (JZ 1) of IRTs in Costa Rica. In our study low species richness in JZ 1 was also evident and was paralleled by a decline in covers of lichens and bryophytes. Microclimatic changes are likely to be greatest close to the ground, but the disproportionate and general impoverishment of the trunk bases (even concerning lichens) may also be related to fire. Indeed, Robertson & Platt (2001) reported direct fire damage to epiphytes up to 1 m above ground.

Concluding remarks

The comparison between the vascular epiphyte flora on IRTs and forest trees showed that numerous taxa decrease in abundance or vanish after isolation. Many of these are typically drought-intolerant, hygrophilous taxa, such as Dryopteridaceae, Grammitidaceae and Hymenophyllaceae. These taxa are partly replaced, if at all, by drought-tolerant, heliophilous species such as *Tillandsia complanata* and *Prosthechea grammatoglossa* (Appendix 2). The most species-rich and diverse IRT sampled in this study had been isolated as recently as 2 years prior to sampling and carried many dead epiphytes, especially pleurothallidinid orchids (Appendix 1, IRT 6).

When isolated from neighboring vegetation, forest trees have higher probabilities of dying than when in the forest interior due to their exposure to adverse environmental conditions (Lovejoy et al. 1986; Lawton & Putz 1988; Kapos et al. 1997). These conditions include higher wind speeds, mean and maximum temperatures, vapor pressure deficit and solar radiation (compare also review by Murcia 1995; Holl 1999). In Panama, edge-interior ratio of trees that died after edges were created was 14:1 (Williams-Linera 1990) and mid-term mortality peaks were found associated with drought events (Laurance et al. 2001). We propose that the same mechanism will affect epiphytes on IRTs.

Guevara et al. (1998) proposed that IRTs function as "stepping stones" for native fauna and "safe sites" for flora, and function as a structurally discontinuous canopy. Our study suggests that IRTs might be "safe sites" for epiphytes at the most in perhumid and arid climates. Here, epiphytic vegetation on IRTs appears to be less reduced compared with the forest than in areas with a moderately seasonal climate such as our study site. It has further been suggested that diversity and composition of epiphytic vegetation on IRTs depends on diaspore influx from adjacent forest vegetation and, consequently, on their distance to these forests (Wolf 1995; Hietz-Seifert et al. 1996; Zimmerman et al. 2000; Nkongmeneck et al. 2002).

In conclusion, we propose that changed epiphyte assemblages on IRTs, compared with those on forest trees, are to a large extent explained by the altered microclimatic conditions on IRTs and their adverse effects on rates of establishment and survival. Reduced availability of suitable niches may further affect epiphyte diversity on IRTs (Sillett et al. 1995; Barthlott et al. 2001). The observed assemblage changes seem to parallel those along the edges of remnant forests. More detailed studies on the epiphyte assemblages of IRTs and along forest edges are necessary to arrive at a better understanding of the unique structure of these assemblages and the processes determining their development.

Chapter

3

DIVERSITY OF DRY FOREST EPIPHYTES ACROSS A GRADIENT OF HUMAN DISTURBANCE IN THE TROPICAL ANDES

WITH S. R. GRADSTEIN

ACCEPTED IN JOURNAL OF VEGETATION SCIENCE

Abstract

Disturbance-effects on dry forest epiphytes are unknown. How are epiphytic assemblages affected by different degrees of human disturbance, and what are the driving forces? Field work was done in a fragmented Interandean dry forest landscape at 2300 m elevation in northern Ecuador. We sampled epiphytic bryophytes and vascular plants on 100 trees of *Acacia macracantha* in five habitats: closed-canopy mixed and pure acacia forest (old secondary), forest edge, young semi-closed secondary woodland, and isolated trees in grasslands. Total species richness in forest edge and on isolated trees was significantly lower than in closed forest types. Species density of vascular epiphytes did not differ significantly between habitat types. Species density of bryophytes, in contrast, was significantly lower in forest edge and on isolated trees than in closed forest. Edge habitat showed greater impoverishment than semi-closed woodland and similar floristic affinity to isolated trees and to closed forest types. Assemblages were significantly nested; habitat types with major disturbance held only subsets of the closed forest assemblages, indicating a gradual reduction in niche availability. Distance to forest had no effects on species density of epiphytes on isolated trees, but species density was closely correlated with crown closure, a measure of canopy integrity. Microclimatic changes are key determinants of the observed impoverishment of epiphyte assemblages following disturbance, and epiphytic cryptogams are sensitive indicators of microclimate and human disturbance in montane dry forests. The substantial impoverishment of edge habitat underlines the need for fragmentation studies on epiphytes elsewhere in the Tropics.

Key words: bioindicator, bryophytes, edge effects, microclimate, nestedness, tropical montane dry forest, vascular epiphytes

INTRODUCTION

Worldwide, dry forests have received much less attention from ecologists than wet and moist forests (Fajardo et al. 2005; Sánchez-Azofeita 2005). The focus on humid sites is particularly obvious in epiphyte research, from which a vast body of knowledge has accumulated (e.g., Kreft et al. 2004). Only a few studies have dealt with epiphyte communities of dry forests. The relatively low diversity of epiphytes in dry forests is presumably the main reason why these forests have received so little attention.

Epiphyte diversity tends to be reduced markedly following disturbance (e.g., Sillett et al. 1995; Acebey et al. 2003; Flores-Palacios & García-Franco 2004; Wolf 2005; Nöske et al., in press); however, there are numerous exceptions to this trend (Hietz-Seifert et al. 1996; Larrea 1997; Nkongmenek et al. 2002; Hietz 2005; Holz & Gradstein 2005a, 2005b). The impact of human disturbance on epiphyte communities of dry forests remains entirely unknown.

For disturbed habitats in humid regions, three mechanisms have been proposed to shape epiphyte diversity: 1) isolation effects, especially constrained dispersal (Cascante-Marín et al. 2006; Pharo & Zartman 2007); 2) reduced structural habitat complexity (Acebey et al. 2003; Krömer & Gradstein 2003; Hietz 1998, 2005); and 3) altered abiotic conditions that increase desiccation stress (Sillett et al. 1995; Krömer & Gradstein 2003; Werner et al. 2005). The first two factors can be presumed to act largely independently of forest type, but this may not be true for the third. By definition, dry tropical forest climates are characterised by the regular occurrence of severe droughts. Dry forest canopies tend to be lower and more open, which results in poor microclimatic stratification (Graham & Andrade 2004). Hence, any dry forest epiphyte should possess effective adaptations to drought and high light levels, which may render them more tolerant to the microclimatic consequences of canopy disruption. This reasoning suggests that disturbance-induced desiccation stress may be less important in dry forests.

In this study we present the first analysis of the effects of human disturbance on epiphyte assemblages of tropical dry forest, and the first direct comparison between disturbance effects on epiphytic bryophytes and vascular plants. Our objectives are to: (1) document patterns of floristic composition and diversity of epiphytes under different disturbance regimes, and (2) explore the driving forces behind these patterns.

METHODS

Study site and sampling

Field work was carried out between January and March 2004 at Bosque Protector Jerusalén in the Interandean Guayllabamba drainage north of Quito. Woodlands in the drainage are strongly dominated by the evergreen *Acacia macracantha*. Primary forests have virtually disappeared from tropical Interandean valleys (Gentry 1992, 1995; Fjeldså 2002), and this reserve harbours one of the most intact Interandean dry forests of Ecuador. The study site is situated on a plateau at 2300–2320 m a.s.l. (S 00° 00', W 078° 21'). It experiences 12 arid months (Guerrón et al. 2005); precipitation averages 530 mm yr^{-1} (INAMHI 1964–1973). The area is characterised by a pronounced valley-mountain breeze typically developing around noon. Fog is uncommon (S. Reyes, pers. comm.). We logged temperature and air humidity 2 m above ground from April 2004–April 2005. Mean annual air temperature was 16.9° C, the coldest month was July (16.3° C), and the warmest months were January and August (17.2° C). Mean daily maximum and minimum temperatures were 24.5° C and 11.8° C. Mean annual relative air humidity was 75.9%, with monthly means ranging from 59.7% (August) to 83.7% (May). Mean daily maximum and minimum air humidity was 93.7% and 45.0%.

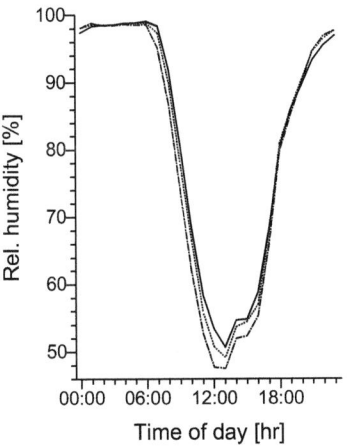

Figure 1. Daily course of relative air humidity at Jerusalén (February–March 2004). Each line represents means of 12 loggers at 2 m height under tree crowns of closed forest (solid line), edge (dotted line) and isolated trees (dashed line). Loggers were set as three triplets for 3-4 days each.

The core area of closed-canopy forest comprises ca. 15 ha of secondary forest surrounded by fallows (semi-closed woodland, scrub) and active pastures. Based on aerial photographs from 1992 and 1976, we estimate the minimum age of the core forest to be ca. 50 yr. At the time of study it was a patchwork of mixed and pure acacia forest stands (habitat types 1 and 2, see below).

We sampled 20 canopy trees of the species *A. macracantha* in each of the following five habitat types:

1. Closed mixed forest.

Along with *A. macracantha*, mixed forest stands included four other tree species: *Buddleja bullata, Caesalpinia spinosa, Mimosa quitensis*, and *Tecoma stans*. Canopy cover was ca. 90–95%. Host tree characteristics are given in Table 1.

2. *Closed* Acacia forest.

This forest type closely resembled mixed forest, but was composed exclusively of *A. macracantha*. Canopy cover was ca. 85–90%. Together with the previous type, this vegetation type will be referred to as closed forest to distinguish them from other vegetation types.

3. Forest edge.

Edge habitat was defined as the first row of trees of closed forest bordering open grasslands. The four edges under study were ca. 450 m length in total, 13–28 yr of age as indicated by aerial photographs, and exposed to the N, S, and E. Edges had been kept open mechanically; the sampled host trees were former forest trees and their architecture closely resembled that of trees in closed forest.

4. Semi-closed secondary acacia woodland.

A continuous patch of open woodland that extended over ca. 5 ha and was bordered by closed forest. This vegetation type had been regenerating from pasture with scattered trees for ca. 20 yr, as indicated by aerial photographs. Because our sampling protocol selected for larger trunks, most of the sampled host trees had presumably established as isolated trees prior to abandonment of use. Canopy cover was ca. 70%.

5. Isolated trees in grasslands.

Isolated trees differed in stature from forest trees, typically having shorter trunks, less strongly inclined branches and dense, low, flat-topped and wide-spreading crowns. Judging from their low stature these were not remnant trees but rather had established in grasslands dominated by *Stipa ichu*. Canopy cover was < 5–10%. Isolated trees were sampled over an area of ca. 100 ha, at distances of 12–2200 m (mean: 537 ± 592 standard deviation [SD]) from closed forest.

Sampling efforts extended over 4 ha or more for all habitat types, except for forest edge. A subset of sample host trees were chosen at random from trees exceeding 25 cm in trunk diameter, with a minimum distance of 25 m between host trees except for the forest edge (minimum = 15 m). Sampling was conducted from the ground and by occasional free-climbing of trees, aided by binoculars. The presence of vascular epiphyte species was recorded for entire host trees, excluding accidental epiphytes (sensu Benzing 1990). Bryophytes were virtually absent from the outer crown and exceedingly scarce in the middle crown. As the middle crown only harboured subsets of inner crown assemblages (F. Werner, unpubl. data), we restricted bryophyte sampling to the inner crown (Zone 3 sensu Johansson 1974). Bryophyte occurrence was highly aggregated and patchy, especially on isolated trees. Random sampling would have yielded a high proportion of 'artificial' zero values. We therefore placed 30 x 20 cm plots on the upper branch sections with highest bryophyte cover. Plots covered branch sections of 30 cm length, and the longitudinal plot axis was kept centered on the top surface line of branches; plots were not moved down the sides of the branches to maximise cover. Bryophytes were removed from one plot per tree and identified in the lab. Although this sampling strategy may have a bias towards late-successional species, Indicator Value Analysis (see below) confirmed that it effectively captured differences in floristic composition between habitat types.

Measured host tree parameters included geographical position (GPS 12, Garmin, Olathe, KS, U.S.A.), height, trunk diameter, and crown closure. Since few trunks (up to the first major ramification) reached breast height, trunk diameter was measured at the point of smallest circumference (usually a few cm below the top of the single trunk). The percentage of crown circumference contacting neighbouring crowns ('crown closure') was estimated to the nearest 5%. Distances of isolated trees to closed forest were measured using ArcGIS 9 (ESRI, Redlands, CA, U.S.A.) and a geo-referenced aerial photograph.

We measured relative air humidity at 2 m height from February 25 – March 13 of 2004, using Hobo Pro data loggers (Onset, Procasset, MA, U.S.A.). Three triplets of loggers (closed forest, forest edge, isolated trees) were moved to new trees every 3–4 days.

Table 1. Host tree characteristics, epiphyte species richness and density throughout habitat types (means ± SD).

	Habitat type				
	Mixed forest (MF)	Acacia forest (AF)	Forest edge (FE)	Semi-closed woodland (SW)	Isolated trees (IT)
Host tree characteristics					
Trunk diameter [cm]	35.81 ± 11.28	36.17 ± 10.67	39.22 ± 9.96	40.70 ± 14.04	43.96 ± 15.68
Tree height [m]	8.01 ± 1.33	6.28 ± 1.16	6.94 ± 1.02	5.42 ± 0.86	5.09 ± 1.37
Crown closure[a] [%]	92.75 ± 7.86	87.25 ± 9.66	49.50 ± 17.73	77.00 ± 11.57	9.38 ± 14.23
Species richness					
S_{obs}	21	19	10	16	12
95% conf. intervals	18.74 – 23.26	15.66 – 22.34	8.68 – 11.32	11.55 – 20.45	9.79 – 14.21
Species density					
Ferns	0.25 ± 0.55	0.10 ± 0.31	0.00 ± 0.00	0.05 ± 0.22	0.00 ± 0.00
Bromeliads	3.20 ± 0.41	3.20 ± 0.41	3.00 ± 0.00	3.30 ± 0.47	3.20 ± 0.62
All vascular plants	3.45 ± 0.76	3.30 ± 0.57	3.00 ± 0.00	3.35 ± 0.59	3.20 ± 0.62
Pleurocarp. mosses	1.95 ± 1.39	1.10 ± 0.79	0.30 ± 0.57	0.70 ± 0.73	0.00 ± 0.00
Acrocarp. mosses	2.00 ± 0.73	1.95 ± 0.94	1.35 ± 0.59	2.00 ± 0.46	0.65 ± 0.59
Liverworts	2.05 ± 1.19	2.10 ± 0.91	1.15 ± 0.59	1.40 ± 1.05	0.50 ± 0.83
All bryophytes	6.00 ± 2.05	5.15 ± 1.50	2.80 ± 0.95	4.10 ± 1.45	1.15 ± 1.14
All epiphytes	9.45 ± 1.99	8.45 ± 1.76	5.80 ± 0.95	7.45 ± 1.28	4.35 ± 1.18

[a] the percentage of crown circumference with direct contact to neighbouring crowns.

Analysis

Richness estimators (Bootstrap, MM-means, Jackknife1, 2; Colwell 2005) indicated adequate sample size, with species accumulation curves levelling off at 10–15 host trees in each habitat type, and estimating < 3 additional species for the most speciose habitat type. Following Colwell et al. (2004), we conducted sample-based rarefaction allowing for open confidence intervals as implemented in EstimateS 7.5 (Colwell 2005).

Between-group differences in species density (the number of epiphyte species per host tree) were analysed with one-way ANCOVA after log transformation, adding trunk diameter as covariate to control for tree size. Where parametric assumptions could not be matched through transformation (vascular epiphytes), the Kruskal-Wallis test was used to analyse differences between groups. Because 20 host trees is a small sample size to test for correlation of species density with distance to forest, we incorporated data of 20 additional isolated trees for this test. These host trees were not used for any other analyses. ANCOVA and correlations were performed using Statistica 6.0 (Statsoft Inc., Tulsa, OK, U.S.A.).

To analyse differences in floristic composition we applied non-metric multidimensional scaling (NMDS) based on Sørensen distance. Validity of a 2-dimensional solution was assessed by means of Monte Carlo randomisations. Simultaneous varimax rotation (Mather 1976) was applied to maximise the loadings of individual variables on the dimensions of the reduced ordination space. NMDS scores were illustrated as double error graphs to facilitate interpretation. Simultaneous plotting of species was done by weighted averaging (Whittaker 1967).

Significance of between-group differences in composition was tested by means of a non-metric multiple response permutation procedure (MRPP), applying a natural weighting factor (n/sum(n)) as recommended by Mielke (1984). NMDS, weighted averaging, and MRPP were performed with PC-Ord 4.25 (McCune & Mefford 1999).

The degree of nestedness of species assemblages provides a measure of non-random patterns in species composition. We used Atmar & Patterson's (1993) 'system temperature' (T), which provides a measure of matrix disorder by relating an observed species x sample matrix to one of the same size and fill that is perfectly nested (ordered, 0°) or completely disordered (100°) (Atmar & Patterson 1993). By applying Monte Carlo randomisations of the observed data matrix (10,000 iterations) we tested for the likelihood that it was randomly generated. Calculations were performed with a program of Atmar & Patterson (1995).

The value of bioindicators is highest when species are truly representative of a group of sites, being unique to that site group (high specificity) and common within it (high fidelity) (McGeoch & Chown 1998). These conditions are combined by Dufrêne & Legendre's (1997) indicator value method (IndVal). We calculated indicator values applying a version of the IndVal index for incidence data (no mean weighting), where: A_{ij} = (Nsites$_{ij}$/Nsites$_i$) where Nsites$_{ij}$ is the number of sites in group j where species i is present, while Nsites$_{i.}$ is the total number of sites occupied by species i. Thus, indicator values were calculated independently for each species and the index was maximal (100%) when a species occupied all sites of a single group and was restricted to that group (Dufrêne & Legendre 1997). Significance of indicator values was tested by means of randomisation (10,000 iterations). Calculations were done with IndVal 2.0 (Dufrêne & Legendre 1997).

Where appropriate, multiple tests of significance were corrected for a table-wide false discovery rate (FDR) of $p < 0.05$ according to the step-up procedure described by Benjamini & Hochberg (1995).

Table 2. Pairwise comparisons of epiphyte species density and floristic composition.

Habitat	Value	Habitat[a]			
		AF	FE	SW	IT
Species density of vascular plants and bryophytes[b]					
MF	P	>0.5	<0.0001	<0.05	<0.0001
AF	P		<0.0001	>0.5	<0.0001
FE	P			<0.05	<0.005
SW	P				<0.0001
Species density of bryophytes[b]					
MF	P	>0.5	<0.0001	>0.1	<0.0001
AF	P		<0.005	>0.5	<0.0001
FE	P			>0.1	<0.0001
SW	P				<0.0001
Floristic composition[c]					
MF	A	0.020	0.148	0.058	0.244
	P	<0.05	<0.0001	<0.0005	<0.0001
AF	A		0.083	0.034	0.215
	P		<0.0001	<0.01	<0.0001
FE	A			0.067	0.108
	P			<0.001	<0.0001
SW	A				0.192
	P				<0.0001

[a] MF = closed mixed forest, AF = closed acacia forest, FE = forest edge, SW = semi-closed secondary woodland, IT = isolated trees.

[b] p-values of ANCOVA post hoc test (Scheffé test, 94 df). * $p < 0.05$. All p-values < 0.05 remained significant after applying the FDR procedure except for floristic composition of AF vs. SF.

[c] A- and p-values yielded by multi-response permutation procedure (MRPP) based on Sørensen distance

RESULTS

Mean relative air humidity in the forest edge and under the crowns of isolated trees was 0.5% and 1.9% respectively below humidity in closed forest. The corresponding hourly Δ RH_{max} was 2.7% at noon (measured 12:00–12:45 h) and 6.5% at 10 h, respectively (Fig. 1). Hourly means differed significantly between habitats during the daytime, from 7–16 h ($p < 0.05$ after FDR; Friedman ANOVA [$n = 12$; $df = 2$]).

We recorded 8 species of vascular epiphytes (5 bromeliads, 3 polypodioid ferns) and 13 species of bryophytes (9 mosses, 4 liverworts). Species richness in forest edge and on isolated trees was significantly lower ($p < 0.05$) than in closed forest types (Table 1).

Individual host trees harboured 2–12 epiphyte species. Mixed forest trees had the highest species density of vascular epiphytes and bryophytes, and isolated trees the lowest (Table 1, Fig. 2). Differences among habitats in total species density were highly significant ($F_{5, 94} = 33.81$, $p < 0.0001$) (Table 2). This, however, was essentially a function of bryophyte density, which differed greatly ($F_{5, 94} = 33.85$, $p < 0.0001$). Species density of vascular epiphytes did not differ significantly between habitats (Kruskal-Wallis test, H [$df = 4$, $n = 100$] = 8.12, $p = 0.090$).

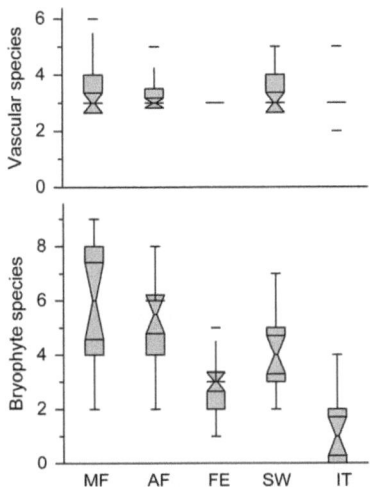

Figure 2. Species density of epiphytes in the five habitat types. Boxes mark the ranges of the middle two quartiles, notches the approximate 95% confidence intervals around the median, whiskers mark adjacent values ($\leq$ 1.5 interquartile range) and dashes mark maxima and minima. MF = closed mixed forest, AF = closed acacia forest, FE = forest edge, SW = semi-closed secondary woodland, IT = isolated trees. The lack of boxes for vascular species in FE and IT is due to data taking only one value (FE) or mostly one value (IT).

Total species density was correlated with crown closure (Spearman rank correlation, r_s = 0.702, n = 100, p < 0.0001) (Fig. 3). Total species density on isolated trees was not correlated negatively with distance to closed forest (r_s = 0.294, n = 40, p = 0.066).

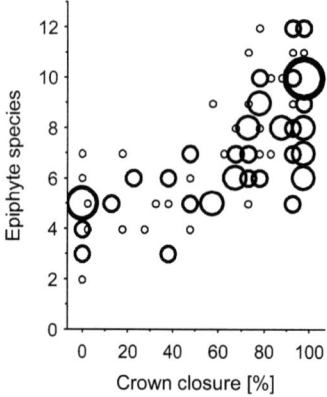

Figure 3. Relation between total epiphyte species density and host tree crown closure (bubble plot). According to their size, bubbles reflect multiple values (maxium 5 host trees).

Floristic dissimilarity was greatest between closed forest habitats and isolated trees, with habitats of intermediate disturbance, semi-closed woodland and forest edge, taking intermediate positions (Fig. 4). Compositional differences were significant in all pair-wise comparisons of habitat types except between the two closed forest types. Forest edge showed similar floristic affinity to isolated trees and to closed forest habitats (Table 2).

All species occurred in the most mature habitats (closed forest), and only two were marginally more frequent in a more open and disturbed habitat (*Orthotrichum diaphanum* and *Racinaea fraseri* in semi-closed woodland). While bromeliads were similarly frequent in secondary woodland or on isolated trees as in closed forest, all bryophytes and ferns were either exclusively found in closed forest or had highest frequencies there, indicating a nested nature of the assemblages (Fig. 4). This was corroborated by a test of nestedness (T = 5.8°, matrix fill = 64%, p < 0.005; Fig. 5).

Six bryophyte species can be designated as reliable indicators of closed forest habitats at p < 0.01: *Cryphaea patens* (indicator value [%] = 38.5), *Fabronia* cf. *jamesonii* (22.5), *Leskea angustata* (31.3), *Syntrichia fragilis* (27.9), and *Microlejeunea globosa* (26.7) for mixed forest, and *Frullania cuencensis* (28.7) for acacia forest. None of the vascular species was identified as an indicator (p < 0.05). Moreover, there were no indicator species for habitat types of intermediate or high disturbance.

DISCUSSION

The small number of epiphytic taxa characterises the study area as a marginal habitat for epiphytes and seems to confirm the importance of humidity as a driver of epiphyte species richness (Gentry & Dodson 1987a; Wolf 1993; Kreft et al. 2004). Other dry neotropical sites are similarly poor: Pedrotti et al. (1988) inventoried vascular plants of an Interandean valley in Bolivia dominated by *A. macracantha* (412 mm prec./an., 2388 m a.s.l.). They found three vascular epiphytes at 2450–2800 m a.s.l, all atmospheric *Tillandsia* spp. Lower sites (2300–2400 m) featured only one of them, *T. bryoides*. Gentry & Dodson (1987a) reported three species of vascular epiphytes from 0.1 ha of seasonally dry lowland forest in coastal Ecuador (804 mm prec./an.) and five additional species from the wider area. Yeaton & Gladstone (1982) reported nine vascular epiphytes from the dry lowlands of Guanacaste in Costa Rica. Bryophyte diversity in dry forests has been very little studied but is generally low and made up of a few drought-tolerant species, mostly of mosses (Gradstein et al. 2001).

Vascular epiphytes did not decrease in species density with increasing disturbance, suggesting that dry forest communities of vascular epiphytes may indeed be relatively disturbance-resilient, but this notion requires confirmation by further studies. Epiphytic bryophytes, on the other hand, responded drastically to increasing disturbance.

The epiphyte flora of semi-closed secondary woodland was distinct from the older, closed forest types, but species richness and density were not significantly reduced (Table 2). Secondary moist forests regenerating after clear-cutting generally show considerably reduced species numbers of epiphytes compared to mature forests (vascular plants: Hietz 1998; Barthlott et al. 2001; Krömer & Gradstein 2003; bryophytes: Acebey et al. 2003; Drehwald 2005; but see Larrea 1997; Holz & Gradstein 2005a, 2005b; Nöske 2005). Along with altered microclimate, this has been attributed to reduced surface area for colonisation, and the lack of late successional substrates on young, small and fast-growing host trees (e.g., Acebey et al. 2003). Because in our study sampling was restricted to larger individuals of a single host tree species, substrate area and quality were not lower in more disturbed habitats relative to closed forest. Moreover, our comparisons are not based on primary forest but on old, closed-canopy secondary forest, which may harbour lower epiphyte diversity than the original primary forest. The recovery of epiphyte diversity in arid secondary woodland may therefore proceed slower than suggested by our results.

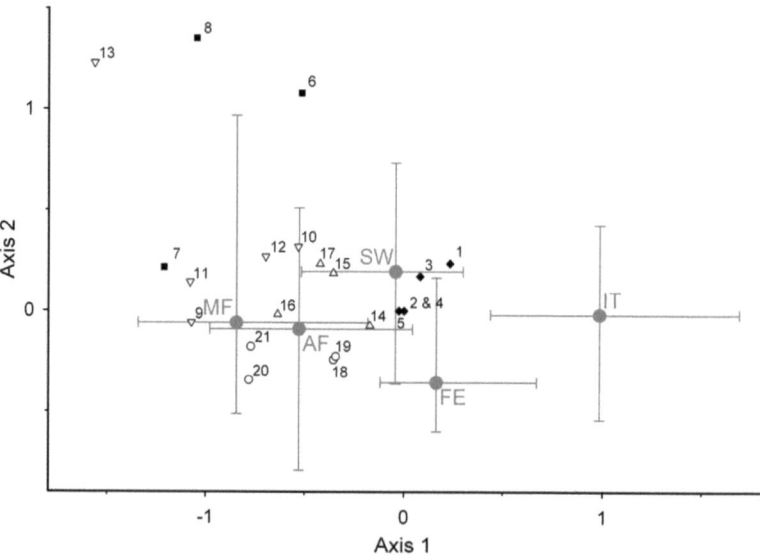

Figure 4. Ordination of samples and species. Double error bar plots (in grey) of NMDS scores (200 runs, 500 iterations; $n = 100$, stress = 0.20) based on Sørensen dissimilarity. Central dots indicate median gravity points of habitat types, whiskers the standard deviations of the mean. Axis 1 reflects species density and crown closure. Epiphyte species (numbered symbols) were added by means of weighted averaging based on the same sample x species matrix. Numbers 1–5 (diamonds) are bromeliads (*Racinaea fraseri, Tillandsia incarnata, T. lajensis, T. recurvata, T. usneoides*), 6–8 (squares) ferns (*Pleopeltis macrocarpa, P. murorum, P. thyssanolepis*), 9–13 (triangles down) pleurocarpous mosses (*Cryphaea patens, Fabronia ciliaris, F.* cf. *jamesonii, Leskea angustata, Orthostichella pentasticha*), 14–17 (triangles up) acrocarpous mosses (*Orthotrichum diaphanum, O. pycnophyllum, Zygodon* sp.*, Syntrichia fragilis*), 18–21 (circles) liverworts (*Frullania cuencensis, Cololejeunaceae minutissima, Lejeunea cardotii, Microlejeunea globosa*). Habitat abbreviations as in Fig. 2.

In terms of species density and floristic composition, isolated trees harboured the most strongly divergent epiphyte assemblage of the five habitats. Because the local epiphyte flora is exclusively composed of wind-dispersed taxa, diaspore rain should decline geometrically with distance from the forest (Madison 1979). As a consequence, reduced diaspore influx may limit the diversity of epiphyte assemblages on isolated trees (Cascante-Marín et al. 2006; Zartman & Nascimento 2006), especially when these are regrowth as in our study. However, we did not find any negative effect of distance to forest on epiphyte species density on isolated trees, suggesting that epiphyte diversity was not constrained by dispersal.

Edge habitat fostered lower total species richness than any other habitat, and in terms of composition was as similar to isolated trees as it was to closed forest (Table 2; Fig. 4). While forest edges are unlikely to receive substantially reduced input of diaspores, they experience increased solar radiation, air temperature, vapour pressure deficit, and wind speed compared to forest interior (Murcia 1995; Laurance 2004; Fig. 1). Although such physical edge effects and their consequences for organisms have received considerable attention in the literature (Harper et al. 2005), the response of epiphytes other than boreal lichens remains poorly known. In the only study dealing with a tropical embryophyte community, edge proximity had only marginal impact on the composition of epiphyllous bryophyte assemblages, and species density was unaffected (Zartman 2003; Zartman & Nascimento 2006).

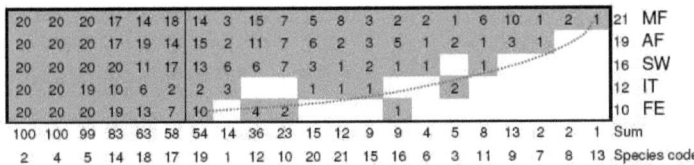

Figure 5. Maximally packed species x habitat type matrix (T = 5.8°, matrix fill = 64%, p = 0.0033). The numbers inside the matrix rectangle reflect the number of host trees for which individual species were registered. Species absences above the boundary line of occurrence for a perfectly nested matrix (dashed) are 'unexpected', as are presences below it. The low number of such unexpected presences and absences is indicative of a high degree of nestedness. When excluding vascular epiphytes, IT and FE take equal positions (T = 4.05°, matrix fill = 63%, p = 0.0040). Species codes as in Fig. 4, habitat abbreviations as in Fig. 2.

Responses to edge exposure clearly depend on the ecological amplitude of the taxa under study. Gradstein (1992) suggested that epiphytic bryophyte assemblages of tropical forest edges generally differ from those of the forest interior by increased representation of drought-tolerant species. At Jerusalén, atmospheric *Tillandsia* species and the widespread moss *Orthotrichum diaphanum* occurred as frequently in forest edge as in forest interior, whereas most other bryophytes and ferns were rare or absent. Species richness and abundance of ferns and bryophytes is closely coupled with humidity (Kessler 2001; Pharo et al. 2005). A similar bias to more xerophilous species as in edge habitat was further exhibited by assemblages of semi-closed secondary woodland and isolated trees. Although virtually all bryophytes decreased in frequency with increasing disturbance, pleurocarpous mosses did so disproportionately (Table 1), a pattern also found by Hylander and

Hedderson (2007) in narrow forest patches in South Africa. The ratio between acrocarpous and pleurocarpous mosses is humidity-dependent, with pleurocarpous mosses becoming relatively scarcer towards drier environments (Frahm 2001; Kürschner 2004). Most importantly, species richness paralleled canopy cover in our study, and species density was strikingly correlated with crown closure (Fig. 3), suggesting a strong influence of microclimate on patterns of floristic composition and diversity.

The microclimate of disturbed Interandean dry forest is markedly altered from that of intact forest (Hohnwald 1999). In closed forest, irradiation is converted into heat at the interface of atmosphere and canopy, preserving moist and cool conditions in the understorey. Disturbed, semi-closed woodlands possess a second 'effective surface' where full sunlight reaches and heats up the ground, which distorts microclimatic stratification (Hohnwald 1999). The resulting differences in air humidity recorded at Jerusalén appear to be moderate (Fig. 1), but even slight differences may be significant for species near their threshold levels of water supply (Murphy & Lugo 1986).

Edge impoverishment appeared disproportional, greatly exceeding impoverishment of semi-closed secondary woodland and approaching that of isolated trees. In addition to growth limitations imposed by physical edge effects, a strong and sudden shift in abiotic conditions with edge creation may have caused high mortality among the established epiphyte flora. For example, in a humid Andean forest, bryophyte cover decreased sharply and mortality of vascular epiphytes was 51% during the first year after host trees were isolated in a new clearing (chapter 7). Structural edge characteristics at Jerusalén may further exacerbate exposure to light and wind. Edges were not closed ('sealed') by regrowth, and while the crowns of isolated and semi-closed woodland trees were generally low and dense, those of edge trees were tall and open (compare Didham & Lawton 1999).

Our study indicates that epiphytic bryophytes are more sensitive to microclimatic changes than vascular plants. Bryophytes lack roots, water storage tissue and an effective cuticle, which links them more closely to their physical environment. Unlike the great majority of vascular plants, most bryophytes are desiccation-tolerant and endure drought instead of avoiding it (Richardson 1981; Frahm 2001). These plants are limited in their ecological distribution by their inability to maintain a positive carbon balance during repeated cycles of wetting and drying (Alpert 2000). For instance, the moss *Grimmia laevigata* must remain hydrated for about 3 h after dawn on a clear day to recoup respiratory losses of carbon after a saturating rainfall the evening before (Alpert & Oechel 1985). The duration of bryophyte turgescence following wetting events (rain, mist, fog) is a function of evapotranspiration and its components (air humidity, temperature, wind turbulence), which are strongly dependent on canopy integrity (Fig.1; Hohnwald 1999; Laurance 2004). Consequently, the duration of turgescence should decrease with crown closure, which may affect the number of bryophyte species able to maintain a positive carbon balance. Support for this interpretation is lent by the high degree of nestedness of epiphyte assemblages at Jerusalén, which was unrelated to other possible non-random causes such as area or distance (Wright et al. 1998).

Conclusions

Contrary to several recent studies on fragmentation effects (see Pharo & Zartman 2007), habitat quality but not dispersal limitation could explain floristic impoverishment in our study system. Specifically, most bryophytes showed great sensitivity to canopy integrity; species such as *Cryphaea patens* provide valuable indicators of microclimate and, locally, of human disturbance. Floristic changes along the disturbance gradient at Jerusalén did not reflect a 'true' ecological gradient in terms of gradual species replacement (turnover). Instead, in more open habitats we observed non-random, xerophyte-biased subsets of the closed forest assemblages, suggesting that species drop out gradually as niche availability diminished. Unlike bryophytes, vascular epiphyte assemblages did not exhibit a significant response to disturbance, which stresses the need for more cross-taxon comparisons. Although isolated trees are considered keystone structures in many aspects (Manning et al. 2006), our results show that their conservation value for epiphyte diversity may be limited (Nöske et al., in press). However, the good recovery of epiphyte diversity that we documented for semi-closed secondary woodland appears to underline the importance of isolated trees as potential nuclei of forest regeneration.

Our results thus confirm the exceptional vulnerability of many epiphytes, particularly bryophytes, to shifts in patterns of humidity, which is the dimension of global climate change most likely to displace or extirpate tropical biota (Benzing 1998; Nadkarni & Solano 2002; Phillips & Malhi 2005). The considerable edge effects documented here raise questions such as (1) how far into the forest interior do single epiphyte taxa respond, and (2) how may taxon-specific amplitudes of response vary with mesoclimate and forest structure? Abundance of certain epiphytic lichens in boreal edges may be reduced 50 m into mature forest (Moen & Jonsson 2003; Esseen 2006). Studying how tropical epiphytes are affected by habitat fragmentation will yield critical information for the conservation of biodiversity in the face of ongoing land conversion and global climate change.

Chapter 4

SPATIAL DISTRIBUTION AND ABUNDANCE OF EPIPHYTES ACROSS A GRADIENT OF HUMAN DISTURBANCE IN AN INTERANDEAN DRY VALLEY, ECUADOR

WITH S. R. GRADSTEIN

Chapter 4 — Spatial distribution and abundance of dry forest epiphytes across a disturbance gradient

ABSTRACT

We studied the effects of disturbance on dry forest epiphyte assemblages. Our objective was to analyse patterns and determinants of epiphyte abundance under different disturbance regimes. Field work was done at Bosque Protector Jerusalén in the Guayllabamba drainage of northern Ecuador (2300 m a.s.l., 530 mm yr^{-1}). Epiphytes on 100 trees of *Acacia macracantha* were sampled in five different habitat types, including closed mixed and pure acacia forest stands, semi-closed secondary acacia woodland, forest edge, and isolated trees in pastures. Host trees were divided into four zones (cf. Johansson 1974). Vascular epiphytes were sampled for entire host trees; macrolichens and bryophytes only for the inner crown. Species density of vascular epiphytes did not differ significantly between crown zones, but their cover was greatest in the middle crown. Vascular epiphytes attained greatest covers on isolated trees, where bryophyte cover was substantially lower than in closed forest. Lichen cover was significantly higher in forest edge than in semi-closed secondary woodland and on isolated trees. Thus, successional patterns along the branch-twig trajectory were peculiar in that bryophytes dominated the latest stage (strong limbs of inner crown) in the forest, while vascular epiphytes (essentially atmospheric bromeliads) attained their greatest abundance in the middle crown. In tropical moist forests, bryophytes typically precede vascular epiphytes, which increase monotonously in biomass towards the inner crown. Distance to forest as a measure of dispersal constraints was not correlated with covers of lichens, bryophytes or vascular epiphytes on isolated trees. In contrast, crown closure, a measure of canopy integrity, was positively correlated with bryophyte cover and, inversely so, vascular epiphyte cover, suggesting that microclimatic changes were a key determinant of epiphyte abundance. Our results suggest that the 'similar gradient hypothesis' proposed by McCune (1993) applies to epiphyte assemblages of tropical dry forest.

Key words: anthropogenic disturbance, biomass, bryophytes, desiccation-tolerance, edge effects, isolated trees, lichens, microclimate, similar gradient hypothesis, tropical montane dry forest, vascular epiphytes

INTRODUCTION

Tropical dry forests potentially cover extensive areas and are likely to gain importance with global warming (Malhi & Phillips 2004; Mayle et al. 2004). Nevertheless, dry forests are among the least known tropical ecosystems (Martinez-Yrizar et al. 2000; Malhi & Phillips 2004). Their study has been severely biased to few countries (Fajardo et al. 2005) and to lowland sites, despite the fact that montane dry forest ecosystems cover a considerable area, foster high species richness and endemism levels, and are severely threatened (Gentry 1992, 1995; Davis et al. 1997; Kessler et al. 2000; López 2003). The negligence of tropical dry forests is particularly obvious in epiphyte research; while a vast body of knowledge on epiphyte communities of moist and wet tropical forests have been accumulated, only few studies have dealt with dry forests, mostly focussing on floristic or physiological aspects (Sanford 1968; Gentry & Dodson 1987a, 1987b; Ibisch 1996; Drehwald 1995, 2005; Graham & Andrade 2004). This is probably related to the fact that epiphytes tend to be less prominent and, particularly, diverse under arid conditions (Gentry & Dodson 1987a, 1987b; Kreft et al. 2004).

Following forest disturbance, epiphyte diversity may be reduced by a number of factors. Degraded and secondary forests usually offer less surface area available for colonisation, with late-successional substrates (e.g., humus and bryophyte mats on old, thick limbs) being particularly scarce (Acebey et al. 2003; Krömer & Gradstein 2003). Distance to propagule sources can affect epiphyte recruitment (Sillett et al. 2000; Hilmo & Såstad 2001; Snäll 2005), and dispersal constraints may therefore limit epiphyte diversity in disturbed habitats (e.g. Hietz-Seifert et al. 1996; Wolf 2005; Zartman & Nascimento 2007). Other authors, who found shade-loving epiphytes underrepresented in disturbed habitats, addressed this primarily to their altered microclimate (Sillett et al. 1995; Krömer & Gradstein 2003; Flores-Palacios & García-Franco 2004). Disturbed habitats are characterised by a fragmented or lower and more open canopy that allows for elevated penetration of light and wind, resulting in disturbed microclimatic stratification and lowered air humidity in lower forest strata (e.g., Malhi & Phillips 2004). In the case of secondary forests recovering from clear-cutting, time since disturbance is an additional factor that merits particular attention, since growth rates of epiphytes tend to be very low compared to terrestrial taxa (e.g., Zotz 1995), and their regeneration invariably lags behind the regeneration of host trees.

Interestingly, there is little consensus between studies regarding the responses of tropical epiphyte assemblages to human disturbance. Epiphyte diversity of disturbed habitats ranges widely, from similar (Larrea 1997; Holz & Gradstein 2005a, 2005b; Nöske 2005) to substantially lower than mature forest (Acebey et al. 2003; Krömer & Gradstein 2003; Drehwald 2005; Werner et al. 2005; Benavides et al. 2006) relative to undisturbed forest. Patterns of epiphyte abundance in disturbed habitats gape even more widely, ranging from significantly lowered (Krömer & Gradstein 2003; Werner et al. 2005; Benavides et al. 2006) to significantly increased (Dunn 2000; Flores-Palacios & García-Franco 2004; Cascante 2006). Moreover, patterns of diversity and abundance can diverge within single studies (Dunn 2000; Cascante 2006), and these apparent discrepancies are not understood.

Because dry forest canopies tend to be more open than those of moist forests, gradients in humidity and exposure are less pronounced (Graham & Andrade 2004), and it may be assumed that the effective adaptations to drought possessed by dry forest epiphytes favour their persistence following forest disturbance. This notion seems supported by reports of limited stratification in dry forest epiphytes (Benzing 1990; Zimmerman & Olmsted 1992). Tropical dry forest epiphytes may therefore respond differently to disturbance than moist and wet forests. With this first study addressing disturbance effects on dry forest epiphytes we aimed at (1) documenting abundance patterns of epiphytes under different types and magnitudes of disturbance and (2) learning about the forces driving epiphyte diversity following disturbance.

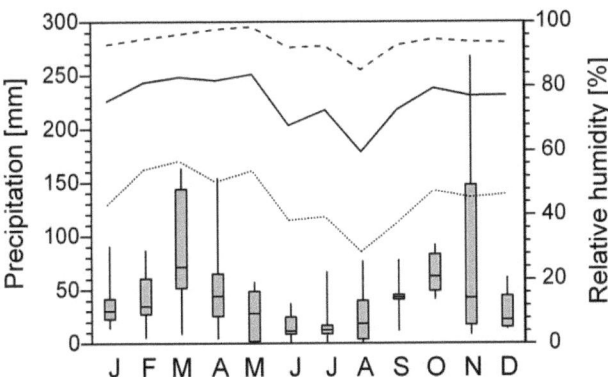

Figure 1. Precipitation and air humidity at Jerusalén. Rain (box-plots; whiskers showing extreme values) for 1963-72 (INAMHI 1964–73). The lines are mean (solid), minimum (dotted) and maximum (dashed) daily air humidity 2 m above ground from April 2004 to April 2005 (original data).

METHODS

Study site and sampling

Field work was carried out by the first author between January and March 2004 at Bosque Protector Jerusalén, a state reserve protecting dry forest, scrub and regeneration pastures in the Interandean Guayllabamba drainage north of Quito. The reserve harbours one of the least disturbed Interandean dry forests of the Ecuadorian Andes. The study site was situated on a plateau at 2300–2320m elevation (S 00° 00', W 078° 21').

Rainfall data from Jerusalén is available for 1963–1972 (INAMHI 1964–1973). Mean annual precipitation was 530 mm with high variability between years (Fig. 1). The area experiences 12 arid months (Guerrón et al. 2005) and is characterized by a pronounced valley-mountain breeze typically picking up around noon, with up to 140 km/h strong winds during the driest season that usually stretches from June–August (Guerrón et al. 2005). Fog is uncommon (S. Reyes, pers. comm.). A detailed habitat description is provided by Werner & Gradstein (in press).

We sampled 20 canopy representatives of the dominant local tree species, *Acacia macracantha*, in five habitat types: closed mixed forest, closed acacia forest, edge of closed forest, semi-closed acacia woodland regenerating from grassland, and isolated trees in pastures (Fig. 2). Information on structure and disturbance history is given by Werner & Gradstein (in press).

Host trees were chosen at random among trees exceeding 25 cm of trunk diameter. Sampling was conducted by climbing and from the ground, aided by a pair of Minox BD 42 x 10 binoculars. Trees were divided into four zones modified after Johansson (1974): trunk (Johansson Zone 1), inner crown (major branches; JZ 3), middle crown (minor branches down to ca. 5 cm of diameter; JZ 4) and outer crown (twigs; JZ 5). This scheme was adjusted to individual trees in such way as to yield three crown zones of equal surface area. Trunks were short (ca. 1 m on average) and therefore invariably had a lower surface area. The presence of vascular epiphyte species was recorded for each of the four zones; accidental epiphytes (sensu Benzing 1990) were recorded separately and are listed in Appendix 3, but were not included in any analysis.

Figure 2. The study site (Bosque Protector Jerusalén). View of the core forest, with the Malchinguí escarpment in the background (a); closed mixed forest (b); forest edge (c); isolated tree (d); atmospheric *Tillandsia usneoides* (left), *T. recurvata* (right) and few juvenile *T. incarnata* (top left) (e).

In the study area, three widespread CAM-metabolizing atmospheric *Tillandsia* spp., most notably *T. recurvata*, were omnipresent and settled in trees at great densities; we estimate larger trees to carry in the range of 1000–10000 genets of *T. recurvata* alone, many of which were interwoven in dense mats. Under such circumstances the count of individuals or 'stands' is not feasible. We therefore estimated surface cover, which, for similar reasons, is the standard measure of abundance for lichens and bryophytes (Gradstein et al. 2003). Vascular plant cover was estimated for each of the Johansson zones. Since abundance of bryophytes and lichens in the middle and outer canopy could not be estimated accurately, their cover was registered only for inner crowns. Cover was estimated to the nearest 5% with two additional steps of 1 and 2% respectively.

As a measure of canopy integrity the percentage of crown circumference contacting neighbouring crowns ('crown closure') was estimated to the nearest 5% for each of the hosts. Using ArcGIS 9 we laid the GPS positions of isolated trees over a geo-referenced aerial photograph to measure the closest distance of each host to mature forest.

Analysis

Between-group differences were analysed with a one-way analysis of covariance (ANCOVA) after log (vascular cover Z3, Z5), arcsine-root (vascular cover Z4, crown means [Z3-5]) and 1/squareroot transformation (bryophyte and lichen cover) respectively. Trunk diameter was added as a covariate to control for tree size, which influences measures of diversity and abundance through substrate age and area available for settlement (Flores-Palacios & García-Franco 2006). Where parametric assumptions could not be matched through transformation, the Kruskal-Wallis test and subsequent Mann-Whitney U-tests were used to test for differences between groups. For the correlation between covers on isolated trees and distance to forest we included 20 additional data points of isolated trees. Vascular epiphyte cover was correlated using the mean of the three crown zones (JZ3–5), omitting the trunk for its locally low suitability and importance as growth site for vascular epiphytes. Analyses were done with Statistica 6.0. Where appropriate, multiple tests of significance were corrected for a table-wide false discovery rate (FDR) of $p < 0.05$ according to the step-up procedure described by Benjamini & Hochberg (1995).

RESULTS

We recorded 31 species of epiphytes. Eleven species of bryophytes were 'truly epiphytic' (holoepiphytic) and two were accidentally epiphytic (sensu Benzing 1990). Eight of the 16 recorded vascular plant species were accidentally epiphytic (Appendix 3). Accidental vascular epiphytes occurred on 21% of trees; the accidental *Capsicum rhomboideum* (Solanaceae) was more frequent than any other vascular species except atmospheric bromeliads (Appendix 3). With a single exception, all accidental vascular epiphytes rooted in deep bark crevices or principal forks filled with dust (clay).

Where present, true vascular epiphyte species showed similar spatial distribution throughout habitats (Appendix 4). They occurred with at least two of the eight local species on all trees sampled (*T. incarnata*, *T. recurvata*). The two tank bromeliads, *Racinaea fraseri* and *Tillandsia lajensis*, occurred mostly in the outer crown, whereas the three small atmospheric *Tillandsia incarnata*, *T. recurvata* and *T. usneoides* were omnipresent in the crown but less common on the trunk (Table 1). Ferns were largely confined to the inner crown where they contributed less than 1% of total vascular cover. Differences in the spatial distribution of epiphyte species density were small between crown zones, but trunks were poorly colonised, reflecting their small surface area (Table 1). Epiphyte species density (the number of species found per tree) did not differ between habitats for any of the crown zones (all $p > 0.1$; $n = 20$; Kruskal-Wallis test).

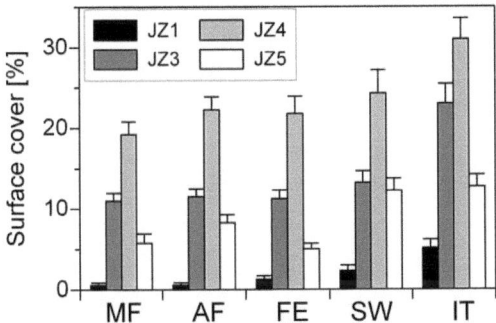

Figure 3. Spatial distribution of vascular epiphyte covers throughout habitat types. Shown are mean covers and standard errors for each Johansson zone (JZ). MF = mature mixed forest, AF = mature acacia forest, FE = mature forest edge, SW = semi-closed woodland, IT = isolated trees.

Across habitat types, the highest cover of vascular epiphytes was found in the middle crown, while the lowest cover was found on trunks. Pairwise comparisons showed significant differences between all Johansson-zones (all $p < 0.0001$; $n = 100$; Wilcoxon test). Throughout Johansson-zones the highest cover was consistently found on isolated trees (Fig. 3). Furthermore, vascular plant cover was significantly higher on isolated trees than in all woodland habitats for trunk and inner crown (Table 2). Lichen cover, on the other hand, was highest in forest edge and lowest in secondary forest (Fig. 4). Bryophyte cover varied widely, from 16.6% ± 9.7 in closed mixed forest to 0.8% ± 1.0 on isolated trees (Fig. 4). Bryophyte cover was similar in semi-closed woodland and closed forest types, but significantly reduced in forest edge (Table 2).

Table 1. Spatial distribution of epiphyte species frequencies (pooled sample).

	JZ 1	JZ 3	JZ 4	JZ 5	Total
Racinaea fraseri (Baker) Spencer & Sm.	–	1	5	11	15
Tillandsia incarnata Kunth	15	99	99	100	100
Tillandsia lajensis André	–	1	1	5	5
Tillandsia recurvata (L.) L.	35	100	100	100	100
Tillandsia usneoides (L.) L.	9	89	99	99	99
Pleopeltis macrocarpa (Bory ex. Willd.) Kaulf.	–	3	1	–	4
Pleopeltis thyssanolepis (Braun ex Klotzsch) E.G.Andrews & Windham	–	2	–	–	2
Pleopeltis murorum (Hook.) A.R. Sm., comb. ined.	–	2	–	–	2
Total richness	3	8	6	5	8
Mean species density	0.59 ± 0.85	2.98 ± 0.49	3.04 ± 0.32	3.15 ± 0.41	3.26 ± 0.58

Bryophyte cover was correlated positively with crown closure ($r_S = 0.658$, $n = 100$, $p < 0.0001$; Fig. 5), whereas vascular plant cover showed a weak negative correlation ($r_S = 0.437$, $n = 100$, $p < 0.0001$; Fig. 5). Lichen cover was not correlated with crown closure (Spearman's $r_S = 0.002$, $n = 100$, $p > 0.5$).

We found no indication of dispersal constraints. Bryophyte cover on isolated trees was not correlated with distance to forest (Spearman's $r_S = -0.093$, $n = 40$, $p = 0.569$), nor were covers of lichens and vascular plants ($r_S = -0.259$, $n = 40$, $p = 0.107$ and $r_S = 0.209$, $n = 40$, $p = 0.195$, respectively).

Table 2. Differences in epiphyte covers throughout habitat types. Pairwise comparisons by means of Scheffé test (df = 5, 94) except for vascular plants in JZ1 (Mann-Whitney U-test; df = 4); p-values that remain significant after FDR are in bold. Habitat abbreviations as in Fig. 3.

	AF	FE	SW	IT
Lichens JZ3 (F = 6.476, p < 0.0001)				
MF	> 0.5	> 0.5	> 0.1	> 0.5
AF		< 0.1	> 0.5	> 0.5
FE			**< 0.001**	**< 0.05**
SF				> 0.5
Bryophytes JZ3 (F = 48.109, p < 0.0001)				
MF	> 0.5	**< 0.001**	> 0.5	**< 0.0001**
AF		**< 0.0001**	> 0.1	**< 0.0001**
FE			**< 0.01**	**< 0.0001**
SF				**< 0.0001**
Vascular plants JZ1 (p < 0.0001)				
MF	> 0.5	> 0.1	< 0.1	**< 0.0001**
AF		> 0.1	< 0.1	**< 0.0001**
FE			> 0.1	**< 0.001**
SF				**< 0.05**
Vascular plants JZ3 (F = 11.307, p < 0.0001)				
MF	> 0.5	> 0.5	> 0.5	**< 0.0001**
AF		> 0.5	> 0.5	**< 0.001**
FE			> 0.5	**< 0.0001**
SF				**< 0.01**
Vascular plants JZ4 (F = 4.433, p < 0.005)				
MF	> 0.5	> 0.5	> 0.5	**< 0.01**
AF		> 0.5	> 0.5	> 0.1
FE			> 0.5	< 0.1
SF				< 0.1
Vascular plants JZ5 (F = 8.2052, p = 0.000002)				
MF	< 0.1	> 0.5	**< 0.01**	**< 0.001**
AF		> 0.1	> 0.1	> 0.1
FE			**< 0.01**	**< 0.001**
SF				> 0.5
Vascular plants JZ3-5 (F = 9.060, p < 0.0001)				
MF	< 0.1	> 0.5	> 0.1	**< 0.0001**
AF		> 0.5	> 0.5	**< 0.01**
FE			> 0.1	**< 0.001**
SF				< 0.1

DISCUSSION

Patterns of epiphyte frequency and abundance

In all habitat types inventoried, lichens substantially exceeded bryophytes in cover, reflecting the contrasting ecological demands of the two groups. While most bryophytes favour shade and demand high humidity (Gradstein et al. 2001), lichens usually require high light levels and frequent desiccation (Sipman & Harris 1989). The great relative importance of bromeliads in the study area, especially of atmospheric *Tillandsia* species, is also characteristic of arid sites (Pedrotti et al. 1988; Benzing 1990; Ibisch 1996); indeed, the highest degree of adaptation to drought among vascular epiphytes has been reported from atmospheric bromeliads (Benzing 1990; Martin 1994).

While species density of vascular epiphytes in moist forests typically increases markedly from the outer towards the inner crown (Johansson 1974; Rudolph et al. 1998), such an accumulation of species with time and substrate age was not observed at Jerusalén (Table 2). This reflects the xerophilous, early successional character of the local bromeliads, which constituted all common vascular species and accounted for the bulk of epiphyte cover. Although cover peaked in the middle crown, small twigs (outer canopy) clearly showed the highest rates of bromeliad establishment as expressed by seedling numbers (see also Bernal et al. 2005), whereas dead individuals of atmospheric *Tillandsia* species were conspicuously common in the inner crown (F. Werner, pers. obs.).

Highest covers of vascular epiphytes (bromeliads) were consistently found on isolated trees. Increased bromeliad abundance or biomass on greatly exposed trees has often been observed (Bartoli et al. 1993; Caldíz & Fernández 1995; Lowman & Linnerooth 1995; Hietz-Seifert et al. 1996; Dunn 2000; Cascante 2006). These trees apparently offer favourable growth conditions for shade-intolerant *Tillandsia* species which are confined to the uppermost canopy stratum in forest (Pittendrigh 1948; Lowman et al. 1999; Flores-Palacios & García-Franco 2004).

Ferns showed a reverse spatial distribution pattern. They were decidedly late-successional and scarce, even near the ground in closed forest and semi-closed woodlands where they exclusively occurred (Werner & Gradstein, in press). This pattern is surprising, since these ferns tend to be markedly xerophilous in other regions. E.g., *Pleopeltis macrocarpa* constituted 98% of all vascular plants on a 2 yr old, exposed terracotta roof in moist montane southeast Ecuador (F. Werner, pers. obs.), and one of few vascular epiphyte species that was well-represented on isolated remnant trees in the same area (Werner et al. 2005). On isolated trees in the wet lowlands of southern Mexico *P. polypodioides* and *P. astrolepis* were characteristic of exposed twigs and small branches where few other epiphytes occurred (Hietz-Seifert et al. 1996; see also Hietz & Briones 1998 for *P. mexiana*), showing that even xerophilous upper canopy epiphytes can be confined to the most protected lower forest strata near the aridity threshold of their ecological amplitude.

Hence, bromeliads pioneered together with fruticose lichens (especially *Teloschistes exilis*). Both groups were eventually replaced by foliose lichens (e.g., *Heterodermia, Parmotremia, Physcia*), bryophytes, and the occasional ferns in closed forest. This pattern differs from moist forests where bryophytes precede vascular plants, which increase monotonously in biomass towards the inner crown (Schimper 1888; Dudgeon 1923; Nadkarni 1984; Freiberg & Freiberg 2000; Gehrig 2005).

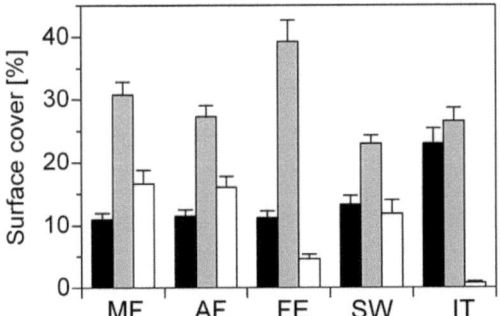

Figure 4. Inner-crown covers of vascular plants (black), lichens (dark grey) and bryophytes (light grey) throughout habit types (means and standard errors). Habitat abbreviations as in Fig. 3.

Processes

Bryophytes were not found at all on almost 40% of isolated trees. Since some of these trees were at a distance of over 2 km from forest, the low bryophyte cover in this habitat may reflect dispersal-limited colonisation. However, bryophyte cover was not correlated negatively with distance to forest, suggesting that dispersal constraints were negligible. Quantity and quality of substrate available for colonization can neither explain the poor representation of bryophytes and ferns in disturbed habitats, given the similar sizes of sampled host trees, and our restriction to one tree species. However, bryophyte cover dropped significantly and sharply with crown closure. Since bryophyte abundance is closely coupled with humidity (Kessler 2001), this indicates a critical influence of microclimatic changes on bryophyte performance following disturbance.

Hohnwald (1999) showed that the microclimate of disturbed Interandean forests differs markedly from closed-canopy forest. In closed forests irradiation is converted into heat at the interface of atmosphere and canopy, preserving moist and cool conditions in the understory. Semi-closed woodlands possess a second 'effective surface' where full sunlight reaches and heats up the ground, which distorts microclimatic stratification (Hohnwald 1999). At Jerusalén, mean relative humidity at 2 m height below the crowns of isolated trees was 1.9% lower than in mature forest (maximum at 10–11 h: 6.5%; Werner and Gradstein, in press), and mean temperature was elevated by 0.2 °C (maximum at 9–10 h: 0.6 °C; n = 12 loggers; F. Werner, unpubl. data). It can be assumed that the resulting increase in evapotranspiration is amplified by elevated velocity and turbulence of wind within the crowns of isolated trees (Flesch & Wilson 1999; Laurance 2004), thus affecting epiphyte performance under the pronounced arid local conditions.

Maybe the most striking outcome of this study was the contrasting response of (atmospheric) *Tillandsia* species versus bryophytes to canopy fragmentation: while the former were significantly more abundant on isolated trees, the latter were much scarcer than in closed forest. Atmospheric *Tillandsia* species are densely covered by highly derived foliar trichomes that allow for effective interception and uptake of moisture from fog and dew. Dew can supply considerable amounts of water: in Mexican dry forest, calculated dew depositions attained 0.72 mm d^{-1} (Andrade 2003). Moreover, dew deposition is greatest during arid times, when daily temperature fluctuations are much larger than during times of rain, which presumably enhances the benefit of dew to plants. Within closed forest, dewfall increases with height and is most frequent and abundant at the interface of vegetation and atmosphere (uppermost canopy) because of the negative nocturnal radiation balance of outer canopy leaves (Nobel 1999; Wilson et al. 1999; Andrade 2003; Dietz et al. 2007). The quantity of dewfall below the canopy is further limited by condensation heat from dewfall that prevents excessive cooling in lower forest strata (Dietz et al. 2007). Unlike in forest trees, the entire crowns of isolated trees form the interface of vegetation and atmosphere, which should consequently receive more dew. Where the soil is not insulated by a canopy (e.g., in the crown periphery of isolated trees), nocturnal heat losses and resulting dew deposition are greatest near the ground (Burckhardt 1963; Bendix 2005), so that dew deposition should be enhanced for plants on isolated trees and in semi-closed woodland, where trees were shorter than in forest edge (Werner & Gradstein, in press). This interpretation is in agreement with the observation that very low (peripheral) branches of isolated trees tended to carry the heaviest loads of atmospheric *Tillandsia* species (Fig. 2).

High dew input is potentially beneficial also to bryophytes, but, near the vegetation–atmosphere interface, comes at a cost. Since wind turbulence and solar radiation are higher near the vegetation–atmosphere interface than in the forest understorey, diurnal vapour pressure deficits (VPD) are elevated, resulting in a shorter duration of surface wetness following wetting events (Dietz et al. 2007). In contrast to atmospheric *Tillandsia* species and most other vascular epiphytes, bryophytes lack effective mechanisms of water loss control or water storage – their strategy is to endure drought instead of avoiding it (Richardson 1981; Proctor & Tuba 2002). Such desiccation-tolerant plants are limited in their ecological distribution by their inability to maintain a positive carbon balance during repeated cycles of wetting and drying (Zotz 1999; Alpert 2000). For instance, the moss *Grimmia laevigata* must remain hydrated for about 3 h after dawn on a clear day to recoup respiratory losses of carbon after a saturating rainfall the evening before (Alpert & Oechel 1985). In other words, when high VPD reduces the length of diurnal turgescence below a (species-specific) threshold, wetting by dew will even result detrimental for bryophytes (see also Csintalan et al. 2000). It is thus likely that under the distinctly arid local conditions bryophytes are largely restricted to lower forest strata because of its low VPD as a consequence of low levels of light and wind turbulence (Laurance 2004). Here, limited evapotranspiratory losses allow for sufficiently long photosynthetic activity after wetting events, namely after saturating rainfalls, when cloud cover limits evaporation from both direct sunlight and valley-mountain breezes. This interpretation is supported by the high degree of nestedness of epiphyte assemblages at

Jerusalén, which closely followed patterns of crown closure for bryophytes (Werner & Gradstein, in press).

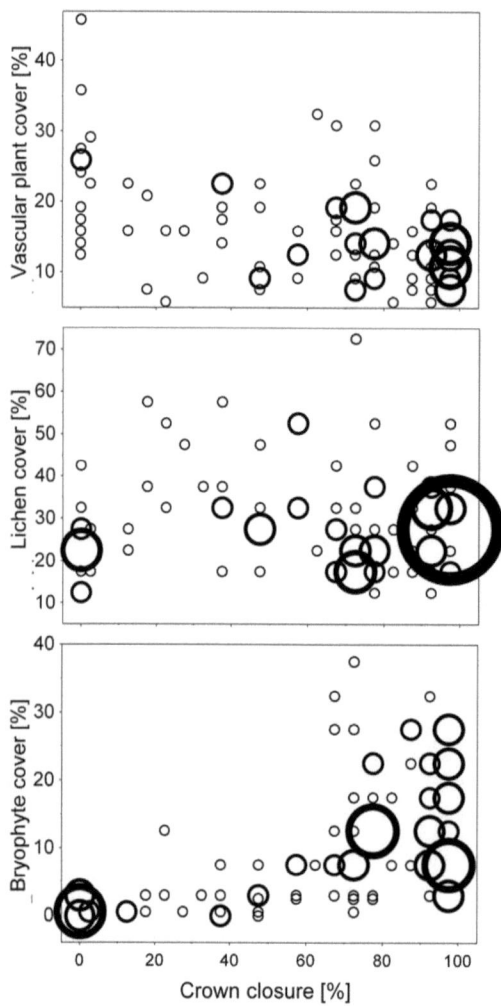

Figure 5. Relation between crown closure and the cover of bryophytes, lichens (inner crown) and vascular plants (crown means) respectively. Bubbles represent 1–10 host trees according to their size. Lines reflect polynomial fit (2nd degree).

The draw-back of the foliar trichomes in atmospheric bromeliads is that they reflect (reduce) light while dry, limiting plant performance in the shady inner crown. In wet state, foliar trichomes inhibit gas exchange and increase vulnerability to photo-injury (Benzing & Renfrow 1971; Benzing et al. 1978; Martin 1994). The persistence of atmospheric bromeliads in the inner crown may further be limited by pathogens. Fungal and bacterial infections are positively correlated with leaf wetting (e.g., Everts & Lacy 1990), and pathogen-induced rotting after prolonged plant-wetting is well-known as a common cause of death in cultivated atmospheric *Tillandsia* (J. Lautner, pers. comm.).

Noteworthily, ferns (3 spp. of *Pleopeltis*) responded similar to bryophytes in that they were restricted to the inner crown, most species-rich and abundant in mixed forest, and entirely absent from forest edge and isolated trees (Appendix 3). These ferns were usually encountered in strongly dehydrated state (curled-up, brittle fronds), but rehydrated readily after rain or watering in the lab, and are most likely desiccation-tolerant as many of their congeners (Hietz & Briones 1998; Schneider et al. 2004; M. Kessler, pers. comm). Unlike bryophytes, ferns can control water losses through an effective cuticle to some extent, but they nevertheless desiccate regularly and quickly after rain events (F. Werner, pers. obs.). It thus appears that in contrast to the strategy of drought-avoidance followed by bromeliads, the strategy of drought-endurance pursued by epiphytic bryophytes and desiccation-tolerant ferns approaches a limit under the aseasonally arid local conditions. Although a similar situation can readily be observed in terrestrial (perennial embryophyte) communities of some subtropical deserts, this does not necessarily hold true for other habitats characterized by excessive drought, for instance inselbergs (Porembski & Barthlott 2000; Proctor & Tuba 2002).

Concluding remarks

Based on studies on epiphytic assemblages in northern temperate conifer forests, McCune (1993) proposed the 'similar gradient hypothesis'. According to this hypothesis, epiphyte taxa are ordered similarly across three different spatial and temporal gradients: (1) vertical differences in composition within a given stand, (2) compositional differences among stands differing in moisture regime but of the same age, and (3) changes in composition through time in a given stand (McCune 1993), which is approximately equivalent to the disturbance gradient under study in terms of environmental growth conditions (microclimate). Our results from Jerusalén (this study; Werner & Gradstein, in press) suggest that this hypothesis also applies to tropical dry forest assemblages. At Jerusalén, inner crown assemblages of isolated trees and young, semi-closed secondary woodland resembled middle crown assemblages of closed, more mature forest. Here, ferns and bryophytes were less abundant, while bromeliads were more abundant than in inner crowns of closed forest. Ferns in the genus *Pleopeltis*, which are early-successional upper canopy dwellers in moist forests, were essentially restricted to inner crowns at Jerusalén. Likewise, the acrocarpous moss *Orthotrichum diaphanum*, which is characteristic of early-successional assemblages in moist temperate broadleaf forests (Drehwald & Preising 1991), was the most frequent bryophyte in inner crowns and even strongly dominated bryophyte assemblages of isolated trees and forest edges (Werner & Gradstein, in press).

The similar gradient hypothesis implies that corresponding changes in growth conditions across gradients, namely the trade-off between availability of moisture and light, are key determinants for the development of epiphytic flora. Our results from Jerusalén support this assumption: both abundance (this study) and diversity (Werner & Gradstein, in press) of epiphytes were unrelated to distance from closed forest as a measure of dispersal limitations but instead correlated with crown closure. Moreover, epiphyte assemblages were strongly nested and aligned along the gradient of canopy integrity (Werner & Gradstein, in press).

While species richness of epiphytes at our study site was marginal, their biomass was not. In fact, local epiphytic biomass may exceed values of many moist forests. For example, 1.5 ha of lowland moist forest in Venezuela harboured no more than 778 individuals of vascular epiphytes (Nieder et al. 2000), and the crowns of canopy trees in a highly diverse montane moist forest in Ecuador only fostered a vascular epiphyte cover of 10% (F. Werner, unpubl. data). We estimate that some of the larger isolated trees at Jerusalén held several dozen kilograms of fresh weight in bromeliads; atmospheric *Tillandsia* species accounted for a major fraction of fresh plant matter available to herbivores, particularly during periods of drought. The role of epiphytes in the functioning of montane dry forest ecosystems remains virtually unexplored and clearly deserves more attention. Moreover, montane dry forests offer unique opportunities for comparisons both with lowland dry forests and with montane moist forests.

Chapter 5

IS THE RESILIENCE OF EPIPHYTE ASSEMBLAGES TO
DISTURBANCE A FUNCTION OF LOCAL CLIMATE?

WITH M. KESSLER & S. R. GRADSTEIN

SUBMITTED

ABSTRACT

The consequences of human disturbance for epiphyte assemblages remain poorly understood, related to a considerable number of potential predictors involved in field studies. Isolated trees do not differ in their degree of physical exposure and therefore offer an excellent model system for the studying of processes shaping epiphyte assemblages following disturbance. We here explore the role of local climate on the diversity of vascular epiphyte assemblages by studying epiphyte assemblages on isolated trees in a tropical montane dry forest landscape in Ecuador, and by analysing literature on epiphyte diversity of isolated trees. In our case study we found species density of vascular epiphytes on isolated trees unchanged, while total species richness was significantly reduced relative to nearby forest. The comparable literature (four additional studies from Ecuador and Mexico) suggests that disturbance effects on species density of vascular epiphytes vary systematically with local climate. Assemblages of moist and moderately seasonal areas appear to experience considerably stronger impoverishment than those of aseasonally wet or distinctly dry areas. We argue that the integrity of the vertical microclimatic gradient is more crucial for the maintenance of epiphyte diversity in moderately seasonal forests than in aseasonally wet or distinctly dry forests.

Key words: diversity, Ecuador, human disturbance, isolated trees, mesoclimate, microclimate, species richness, tropical dry forest, vascular epiphytes

Introduction

There is much discussion in conservation biology about the extent to which anthropogenic habitats can allow the persistence and migration of primary forest organisms (Putz et al. 2001; Laurance 2006; Gardner et al. 2007). Tropical forests continue to be converted, degraded and fragmented at high pace (Laurance & Peres 2006), and the challenge of global climate change will require species migrations of unprecedented magnitude (Bush 2002; Malhi & Phillips 2004). Understanding how alteration and fragmentation of their primary habitat is perceived by organisms is therefore of great importance for conservation (Gascon et al. 1999; Laurance 2006).

Across-lifeform comparisons suggest that epiphytes are more sensitive to human disturbance than terrestrial plants (King & Chapman 1983; Hickey 1994; Turner et al. 1994). However, case studies in secondary and degraded forests have yielded a wide range of different responses of epiphyte assemblages. Whereas some found them impoverished (Barthlott et al. 2001; Krömer & Gradstein 2003; Wolf 2005), others found them as diverse as those of undisturbed forests (Hietz-Seifert et al. 1996; Larrea 1997). The response of single taxa, for instance bromeliads, varied from significantly less species-rich in some studies (Krömer & Gradstein 2003) to considerably more species-rich in others (Barthlott et al. 2001). This diverging of patterns, however, is little surprising. Non-primary forests differ greatly between each other in age and structure, and in the degree of structural differences from their respective potential vegetation. Differences of potential relevance for epiphytes include local climate (mesoclimate), type and degree of disturbance, phorophyte age, stand structure (e.g., height, closure, canopy openness) and resulting microclimate, substrate quality and quantity (e.g., availability of large, horizontal limbs, humus, bryophyte accumulation), and dispersal limitations. Many of these partly correlated factors are difficult to quantify, and information on few of them has been provided, much less analysed, in papers addressing the epiphyte assemblages of non-primary habitats. This circumstance has hampered comparisons, rendering disturbance effects poorly understood and unpredictable.

Scattered trees isolated in an anthropogenic land use matrix (hereafter termed 'ITs') constitute keystone structures that offer refuge, enhance connectivity, and provide nuclei of regeneration (Janzen 1988; Hietz 2005; Wolf 2005; Manning et al. 2006; Zahawi & Augspurger 2006). Moreover, they offer an excellent model system for the studying of human disturbance effects on epiphytes. ITs can be viewed as a 'discontinuous canopy'

(Guevara et al. 1998) or the smallest possible forest fragment (Williams-Linera et al. 1995), being exposed to maximum edge effects and constrained colonisation. Unlike degraded or secondary forests, ITs do not differ between sites in the relative degree of their physical exposure; neither do they differ from undisturbed forest in important phorophyte parameters (e.g., age, surface area, bark characteristics) unless biased in size or taxonomic composition.

Many authors have stressed the importance of microclimatic changes as the principal driver of epiphyte diversity losses and community changes following disturbance (e.g., Hietz 1998, 2005; Barthlott et al. 2001; Nöske et al., in press). Epiphytes are known for their high and fine-tuned humidity requirements, being linked more closely to the atmosphere than terrestrial biota (e.g., Schimper 1888; Benzing 1998; Kreft et al. 2004). Werner et al. (2005) suggested that the relative degree to which epiphyte assemblages impoverish in species may be linked with local climate. They contended that in perhumid regions epiphyte diversity on ITs may be not as strongly affected by desiccation stress as in areas of intermediate humidity, where a large share of the epiphytic flora depends on the maintenance of a mesic microclimate by intact forest canopies during dry periods. In different arid regions of the Neotropics we observed that ITs often harbour a relatively intact epiphytic flora. Since dry forest canopies are usually low and open and cannot provide pronounced microclimatic buffering (Zimmerman & Olmsted 1992; Graham & Andrade 2004), it may be assumed that their epiphytes must be sufficiently well-adapted to persist drought even under increased exposure as experienced on ITs.

The aims of the present study were (1) to test the hypothesis that the species density of dry forest epiphytes remains relatively unaffected by human disturbance by conducting a case study on ITs, and (2) to explore the general hypothesis that the relative impoverishment of epiphyte assemblages is a function of mesoclimate, by analysing available studies on diversity of epiphytes on ITs.

METHODS

Dry forest case study

Field work was carried out during January – March 2004 at Bosque Protector Jerusalén in the Interandean Guayllabamba drainage north of Quito, on a plateau at 2,300 m a.s.l. (S 00° 00', W 078° 21'). The area experiences 12 arid months (Guerrón et al. 2005); mean annual precipitation is 530 mm (INAMHI 1964–73); fog is uncommon (S. Reyes, pers. comm.). A detailed site description is provided by Werner & Gradstein (in press).

We sampled 40 canopy trees of *Acacia macracantha* in forest and isolated in pastures, respectively. Isolated trees were regrowth and sampled at distances of 12–2200 m from mature forest. Host trees were randomly chosen among trees exceeding 25 cm of trunk diameter. Sampling was conducted from the ground and by climbing, aided by binoculars.

Using EstimateS 7.5 (Colwell 2005), we did sample-based rarefaction ('Mao Tau'; 1,000 iterations). Between-group differences were analysed with one-way analysis of covariance (ANCOVA), adding trunk diameter as covariate to control for tree size. Where parametric assumptions could not be matched through transformation, the Mann-Whitney U-test was used instead. Significance-testing was done with Statistica 6.0 (Statsoft Inc., Tulsa, OK, U.S.A.).

Literature survey

We compiled all available literature on epiphyte assemblages on ITs. Where possible we requested additional information from the respective authors.

An index of aridity was calculated for the sites of those studies allowing meaningful comparisons (e.g., comparable size, number and composition of phorophytes, sufficient site information), dividing monthly precipitation by monthly potential evapotranspiration (PET) (UNEP 1992). PET was calculated following the Thornthwaite method (Thornthwaite & Mather 1957) as:

$$TH = 3.65 \times 10^{-4} d(t) I^{-1} [10 T(t)]^a$$

where *TH* results from the monthly Thornthwaite calculation, $d(t)$ is the total monthly daylight in hours, and $T(t)$ is the monthly mean temperature. The two variables I and a are empirically derived functions of mean annual temperature.

Several studies of relevance do not provide comparable data and were excluded from analysis. Gonçalvez & Waechter (2003) found considerable orchid diversity on ITs but did not study forest trees. In the case of other studies (Williams-Linera et al. 1995; Flores-Palacios & García Franco 2001; Nkongmenek et al. 2002; Flores-Palacios & García Franco 2006), comparisons are impeded by major differences in sampling area between treatments. A study by Hietz (2005) on coffee shade trees was not included because trees were too densely clustered to be considered isolated, with individual trees making crown contact (P. Hietz, pers. comm.). The reduced species density of bromeliads on ITs found by Dunn (2000) and Cascante-Marín et al. (2006) was closely related to smaller phorophyte size. Moreover, bromeliads only constitute a small fraction of the local epiphytic floras and are known to be one of the most resilient and xerophytic taxa (Benzing 1990), suggesting that they are not representative for epiphyte communities as a whole.

Our analyses included two spatial levels of diversity: (1) species density, here defined as the number of species on one (most studies) or few trees (Larrea 1997) of comparable size, and (2) total species richness (the total number of species observed or estimated for a study). One study (Hietz-Seifert et al. 1996) provides results of significance testing for species density and richness, but no mean values. We therefore calculated these values from given linear regression equations (species density vs. DBH) at DBH 0.5 m, and by means of individual-based rarefaction at the minimum frequency of species records ($n = 204$; 1,000 iterations), respectively.

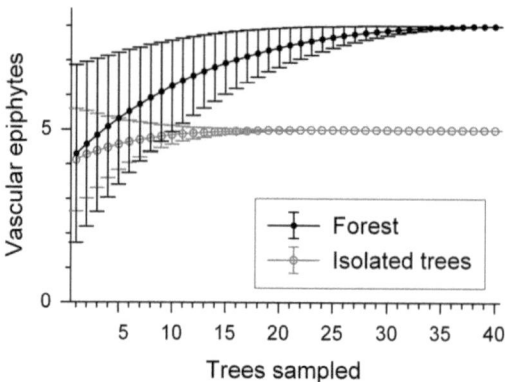

Figure 1. Sample-based rarefaction curves of total vascular epiphyte species richness at Jerusalén Whiskers mark open 95% confidence intervals.

Results

Dry forest case study

Mean phorophyte stem diameter was 36.0 cm ± 10.8 SD in forest trees and 43.2 cm ± 12.6 in ITs respectively; tree height was 7.2 m ± 1.5 in forest trees and 5.1 m ± 1.2 in ITs.

We recorded 8 species of vascular epiphytes (*Racinaea fraseri*, *Tillandsia incarnata*, *T. lajensis*, *T. recurvata*, *T. usneoides* [Bromeliaceae], *Pleopeltois macrocarpa*, *Polypodium murorum* and *P. thysanolepis* [Polypodiaceae]). Single phorophytes held 2–5 (ITs) and 3–6 epiphyte species (forest), respectively. Species density on ITs was not reduced significantly for bromeliads, ferns, or all vascular epiphytes combined (Table 1). However, total species richness on ITs was reduced significantly due to the absence of the three fern species ($p < 0.05$; Fig. 1). Species density on ITs at closer distance to forest (248 m ± 203) was slightly lower (3.00 ± 0.33 species) than for trees at greater distance (1220 m ± 498; 3.45 ± 0.69 species). This difference was not significant (U-test; $p > 0.05$).

Table 1. Species density of different epiphyte groups on 40 trees in closed dry forest and 40 isolated trees in an adjacent agricultural area.

	Forest trees		Isolated trees			
	Mean	SD	Mean	SD	F	p [1]
Ferns	0.18 ±	0.45	0.00 ±	0.00	–	0.25
Bromeliads	3.20 ±	0.41	3.23 ±	0.58	0.04	0.96
Vascular plants	3.38 ±	0.67	3.23 ±	0.58	0.61	0.55

[1] assessed via ANCOVA, except for species density of ferns (Mann-Whitney U-test).

Literature survey

We found four studies providing quantitative, methodologically comparable data on richness of vascular epiphytes on ITs relative to adjacent natural forests (Table 2). Responses of vascular epiphyte assemblages on ITs differed widely between studies (Fig. 2, Fig. 3). In two perhumid areas of SE-Mexico and NE-Ecuador, species density on isolated trees was not significantly lower than on forest trees of similar size (Hietz-Seifert et al. 1996; Larrea 1997). Total species richness and diversity indices (Simpson diversity, Shannon-Weaner entropy) analysed for one of these sites were neither lower (Larrea 1997); rarefied total species richness at the second perhumid site (Hietz-Seifert et al. 1996) was even significantly higher on ITs than on forest trees. At a slightly seasonal moist forest in SE-Ecuador, species density on ITs was > 80% lower than on forest trees (Werner et al.

2005; Nöske et al. in press). Total species richness as estimated with Jack-knife 1 was reduced by 71% relative to the forest (Nöske et al. in press). This highly diverse site regularly experiences short but pronounced dry spells induced by foehn winds (Emck 2007). For inner crowns of ITs at a yet drier and more seasonal site in E-Mexico, Flores-Palacios & García-Franco (2004) found moderate lower species density and total species richness than on forest trees.

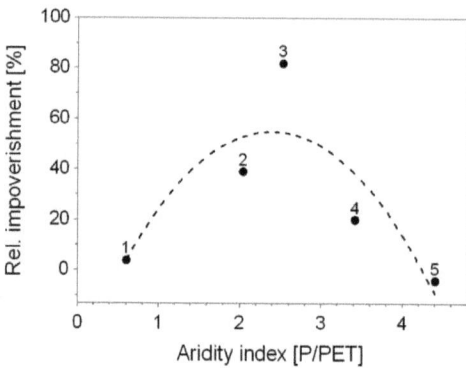

Figure 2. Humidity dependence of species density declines in vascular epiphytes on isolated trees relative to trees in natural forest. Case studies: (1) Jerusalén (this study); (2) Flores-Palacios & García-Franco 2004; (3) Werner et al. 2005; (4) Hietz-Seifert et al. 1996; (5) Larrea 1997. The stippled line reflects polynomial fit (2nd degree).

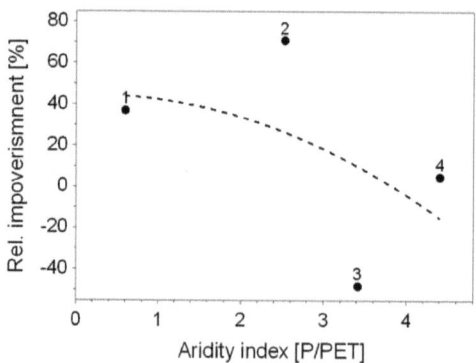

Figure 3. Humidity dependence of total species richness declines in vascular epiphytes on isolated trees relative to trees in natural forest. Case studies: (1) Jerusalén (this study); (2) Flores-Palacios & García-Franco 2004; (3) Werner et al. 2005; (4) Hietz-Seifert et al. 1996; (5) Larrea 1997. The stippled line reflects polynomial fit (2nd degree).

DISCUSSION

In our case study of epiphytic assemblages in the dry forest landscape of Jerusalén we recorded no more than eight epiphyte species, which underpins the paramount importance of humidity for epiphyte diversity (Kreft et al. 2004). Species density of vascular epiphytes was not reduced on ITs, whereas total species richness was reduced significantly. Thus, our hypothesis that epiphyte communities of dry forest are little affected by human disturbance held true for species density, but not for total species richness. The reduction in species richness was due to the absence of three ferns from the sampled ITs. None of these ferns was common in the forest, (2–5 individuals each; chapter 4), where they were restricted to thick moss-covered branches. Since none of the (bark-dwelling) bromeliad species was reduced in abundance on ITs, we attribute the absence of ferns partly to the scarcity of bryophyte mats on ITs (Gehrig 2005; F. Werner, unpubl. data). Clearly, larger distance from forest did not have adverse affects on species density on ITs, suggesting that dispersal constraints were of minor relevance in shaping epiphyte assemblages. However, this conclusion cannot be drawn for ferns, which were entirely absent from ITs. Because all ferns were rare in the forest, particularly regarding fertile individuals, diaspore availability may well be an issue for the colonization of widely scattered ITs, especially where these form regrowth as at Jerusalén.

The literature data suggests that, as predicted, the response of vascular epiphyte assemblages to disturbance does indeed vary systematically with mesoclimate. A strong, significant decrease in species density on ITs was observed only in the two studies with intermediate, slightly seasonal conditions, whereas both the perhumid and arid sites were unaffected (Fig. 2, Table 2). Species density remained high under aseasonally wet conditions where desiccation stress should be limited even on ITs. Thus, continuous moisture supplies appear to facilitate suitable growth conditions for a large number of epiphyte species on ITs, even though post-disturbance turnover may be considerable. In contrast, greatly decreased species numbers appear to result under moist and slightly seasonal conditions, where the bulk of taxa depend on the protection of an intact canopy during periods of drought. Finally, in dry forests, disturbance appears to result in relatively low decreases in species numbers where the vertical microclimate gradient is poorly developed even in intact stands (Graham & Andrade 2004). The effective adaptations to drought which are essential for dry forest epiphytes should greatly promote the persistence

Table 2. Site characteristics and relative impoverishment of vascular epiphytes on isolated trees. Diversity measures reflect proportions of forest values.

Study	Locality	Canopy height [m]	Isolation [yr]	Alt. [m]	Fog	T_{mean} [°C]	$P yr^{-1}$ [mm]	Aridity index	Sampling (no. ITs/ forest trees)	Effect of forest distance	Total species number[a]	Species richness [%][b]	Species density [%][b]
Flores-Palacios & García-Franco (2004)	E-Mexico	30-45[c]	85	1550	common	14.1	1650	2.04	inner crowns (5/5)	n.s.	35		0.61 **
Hietz-Seifert et al. (1996)	SE-Mexico	30-35	28	100	rare	24.5	4700	3.41	entire hosts (38/127)	**	83	1.48[d] *	0.8 n.s.
Larrea (1997)	NE-Ecuador	≤35	5	2200	Un-common	15.2	3500[e]	4.40	entire hosts (30/30)	–	155	0.95 –	1.04 –
Werner et al. (2005)	SE-Ecuador	20	10-30	2000	rare seas.	15.6	2200	2.53	entire hosts (15/6)	n.s.[f]	253	0.29[g] *	0.18 **
This study	N-Ecuador	5-8	n.a.	2300	common	16.9	530	0.60	(40/40)	n.s.	8	0.63 *	0.96 n.s.

[a] values for entire study (both forest trees and ITs).
[b] ratio of means (IT/forest sample) and results of significance-testing.
[c] from Flores-Palacios and García-Franco (2006).
[d] value calculated after Hietz-Seifert et al. (1996).
[e] H. Greeney, pers. comm.
[f] Werner, unpubl. data.
[g] from Nöske et al. (in press).
* $P < 0.05$, ** $P < 0.01$.

of these species on ITs. This conclusion is supported by the paucity of understorey specialists recorded for tropical dry forests. For instance, in coastal Ecuador, Gentry & Dodson (1987a) found 41 species of understorey specialists among vascular epiphytes in a wet forest, but none in a semi-deciduous moist forest. Corresponding patterns have been found in montane E-Mexico by Hietz & Hietz-Seifert (1995a, 1995b).

Factors other than mesoclimate did not help explain the observed patterns of species density. ITs were isolated in pastures in all studies. Phorophyte size was accounted for (Hietz-Seifert et al. 1996) or similar between treatments in all of the studies of comparison, and phorophyte species composition was controlled for in all but possibly one study where entire tree assemblages were sampled at random (Hietz-Seifert et al. 1996). ITs which established after forest clearance may be expected to be more impoverished in epiphytes than remnant trees; however, species density and richness on ITs were unchanged and only moderately lowered, respectively, at the only site where ITs were regrowth (Jerusalén). Time since isolation of ITs or distance to forest as a measure of dispersal limitations are also possible predictors, but were neither related to patterns of impoverishment (Table 2). Distance to forest had significant adverse effects on species density in one study (Hietz-Seifert et al. 1996) but not in the two others where such data was provided (Flores-Palacios & García-Franco 2004, Jerusalén [this study]). Moreover, in the study where impoverishment was most pronounced (Werner et al. 2005), distance to forest neither had a significant effect (Spearman-test: $n = 15$, $r_s = -0.26$, $p > 0.1$; F. Werner, unpubl. data).

A hump-shaped mesoclimate-dependency was apparent in species density, but not in total species richness (Fig. 3). This was especially due to the increased richness found on ITs in one study (Hietz-Seifert et al. 1996). Although phorophyte size-corrected species density on ITs was slightly below forest levels, rarefied total species richness was significantly higher. Here, ITs exceeded forest trees considerably in mean size. Large, old trees are known to harbour the bulk of epiphyte diversity as they offer the most diverse microsites, including especially late-successional substrates such as old mats on thick limbs, crotches, decaying branches or knotholes (Hietz & Hietz-Seifert 1995b; Wolf 2005). A better representation of late-successional substrates by the IT sample is exemplified by patterns of occurrence in hemiepiphytic figs (*Ficus* spp.), which systematically exploit late-successional niches (Laman 1995). Epiphytic figs were represented by only one species in the forest but by six species on ITs, which was not paralleled by a general increase in animal-dispersed epiphyte taxa on ITs. Incomplete sampling as evident in the study from a substantial number of uniques (i.e., singular species records) is clearly unavoidable when dealing with diverse tropical communities, and may result in skewed results where sampling biases cannot be avoided. Therefore, the elevated total species richness on ITs documented by Hietz-Seifert et al. (1996) and the pattern displayed by Fig. 3 should be interpreted with caution.

Patterns in total species richness may indeed be less closely linked to mesoclimate than in species density. Throughout the range of mesoclimates, ITs offer a reduced spectrum of microhabitats, lacking the damp, dark micro-sites found in the interior of closed forests. Especially the most fragile, shade- or moisture-demanding taxa are unlikely to flourish in the multiple edge environment of ITs. This notion is supported by reduced floristic between-site heterogeneity found on ITs in some studies (Larrea 1997; Flores-Palacios & García-Franco 2004). Moreover, a 'true' size and identity of epiphyte communities will often result indeterminable. Especially montane tropical habitats are characterised by high beta-diversity at small spatial scales (Krömer et al. 2005). High rates of species turnover are usually coupled with large shares of rare species, many of which constitute sink populations. Hence, observed (or estimated) total species richness on ITs will almost inevitably be most sensitive to community structure, spatial scale and completeness of a community census, and thus study size. From our own experience, virtually any species may eventually turn up in optimal, azonal microsites on ITs (e.g., knotholes, bromeliad bases). Consequently, the interpretation of ratios in total species richness is less straight-forward than in species density, which may have blurred the patterns underlying Fig. 3.

Conclusions

Our results support the notion of a general relationship between mesoclimate and disturbance effects on species density of vascular epiphytes. Of course, the theoretical relationship between mesoclimate and relative species impoverishment is unlikely to be as symmetrical and plain as suggested by the fitted line in Fig. 2. However, we would expect maximum impoverishment mesoclimatically near to the study site of Werner et al. (2005). Here, 50% of well-established vascular epiphytes tagged in lower forest strata died within one year after their host trees were isolated in a fresh clearing (chapter 7).

Based on a limited number of comparable studies, the proposed hump-shaped relationship (Fig. 2) is still somewhat speculative, and factors other than microclimate such as dispersal limitations may locally be important co-determinants for epiphytic diversity on ITs (Cascante-Marín et al. 2006). We hope that this paper may help to stimulate research on the climate-dependency of patterns and processes in epiphyte communities affected by human disturbance. Advances in this field would have important implications for conservation planning and management, as well as for the prediction of global climate change effects on the diversity and functionality of epiphyte communities (Benzing 1998; Hietz 1998; Nadkarni & Solano 2002; Malhi & Phillips 2004).

Chapter 6

SEEDLING ESTABLISHMENT OF VASCULAR EPIPHYTES ON ISOLATED AND ENCLOSED FOREST TREES IN AN ANDEAN LANDSCAPE, ECUADOR

WITH S. R. GRADSTEIN

ACCEPTED IN BIODIVERSITY AND CONSERVATION

ABSTRACT

The impacts of human disturbance on colonization dynamics of vascular epiphytes are poorly known. We studied abundance, diversity and floristic composition of epiphyte seedlings establishing on isolated and adjacent forest trees in a tropical montane landscape. All vascular epiphytes were removed from plots on the trunk bases of *Piptocoma discolor*. Newly-established epiphyte seedlings were recorded after two years, and their survival after another year. Seedling density and the density of taxa (families and genera, respectively) per plot were significantly lower on isolated trees relative to forest trees, as was the rarefied total number of taxa. Seedling assemblages on trunks of forest trees were dominated by hygrophilous understorey ferns, whereas assemblages on isolated trees were dominated by xerophilous canopy taxa. Colonization probability was significantly higher for plots closer to forest but not for plots with greater canopy or bryophyte cover. Mortality on isolated trees was significantly higher for hygrophilous than for xerophilous taxa. Our results show that altered recruitment can explain the long-term impoverishment of post-juvenile epiphyte assemblages on isolated remnant trees. We attribute altered recruitment to a combination of dispersal constraints and the harsher microclimate that was documented by measurements of temperature and humidity. Although isolated trees in anthropogenic landscapes are key structures for the maintenance of forest biodiversity in many aspects, our results show that their value for the conservation of epiphytes can be limited. We suggest that abiotic seedling requirements will increasingly constitute a bottleneck for the persistence of vascular epiphytes in the face of ongoing habitat alteration and atmospheric warming.

Key words: climate change, diversity, desiccation stress, human disturbance, microclimate, recruitment, species richness, tropical montane moist forest

INTRODUCTION

Vascular epiphytes are a major element of tropical forest structure and biodiversity. They are characterized by traits that may reduce their resilience compared to terrestrial herbs: long generation cycles and sensitivity to atmospheric conditions. These traits are related to resource limitations (nutrients and water) that appear to characterize the epiphytic habitat (Benzing 1998; Zotz & Hietz 2001).

Not surprisingly, the majority of studies on vascular epiphyte assemblages have found pronounced adverse effects of habitat alteration, such as reduced diversity or floristic composition biased towards generalists and xerophytes (Barthlott et al. 2001; Krömer & Gradstein 2003; Flores-Palacios & García-Franco 2004; Hietz 2005; Werner et al. 2005). In addition to increased desiccation stress (Hietz 2005), community changes following disturbance have repeatedly been attributed to constrained dispersal (Wolf 2005) and, in secondary forests recovering from clear-cutting, reduced quality of substrate available for colonization (e.g., lack of bryophyte mats; Krömer & Gradstein 2003). The importance of dispersal constraints has been emphasized particularly in the ample literature on non-vascular epiphytes (Pharo & Zartman 2007). Where epiphyte assemblages survive initial disturbance, such as in moderately logged forest, forest fragments or on remnant trees, impoverishment should proceed gradually due to lowered survival of well-established plants or reduced establishment. The latter may be caused by limited diaspore rain or seedling performance.

Many studies have addressed the performance of different life-stages in trees (see e.g., Hubbell 2001), and the critical role of the early establishment phase for future community composition is well-established (Lieberman 1996). As with trees, epiphyte dispersal can be limiting and seedling mortality is high, suggesting that establishment also plays a critical role for epiphytes. Studies on early epiphyte life-stages incur particular difficulties, including minute and delicate diaspores, lack of keys to identify seedlings, destructive access, and extreme spatial heterogeneity of the epiphytic habitat. Our present understanding of early life-stages of vascular epiphytes and their role in population dynamics is largely based on few species of hemi-epiphytic figs, orchids, and bromeliads in intact forest or green-houses (Benzing 1978; Larson 1992; Laman 1995; Castro Hernández et al. 1999; Doyle 2000; Zotz 2004b). Consequently, the mechanisms by which aspects of human disturbance affect epiphytes are unclear.

Isolated trees in an anthropogenic land use matrix (hereafter referred to as 'ITs') constitute keystone structures that offer refuge, enhance connectivity, and provide nuclei of regeneration (Janzen 1988; Wolf 2005; Manning et al. 2006; Zahawi & Augspurger 2006). Moreover, they offer an excellent model system for the studying of human disturbance effects on epiphytes, being exposed to multiple edge effects, constrained colonization, and being easily replicable.

After 10–30 years of isolation in pastures, isolated remnant trees in Ecuador harboured post-seedling epiphyte assemblages that were substantially impoverished in terms of abundance and species richness, and strongly biased to canopy taxa (Werner et al. 2005). The purpose of this study was to examine the role of establishment in such long-term community alterations at the same study site, using an experimental approach. Specifically, we examined: (1) if seedling establishment is reduced on ITs, (2) to what extent colonization patterns mirror the biased floristic composition observed on ITs after prolonged isolation, and (3) potential factors for altered establishment.

METHODS

Study site

Field work was done at 1800–2000 m a.s.l. in the surroundings of Estación Científica San Francisco (ECSF) near Podocarpus National Park in southeast Ecuador (3° 58' S, 79° 04' W). The natural vegetation of slopes and ravines is moist montane forest with a canopy height of 15–25 m (Homeier et al., in press). The area fosters a remarkably rich epiphytic and terrestrial flora (Homeier and Werner, in press; Lehnert et al., in press).

Mean annual temperature at 1950 m is 15.5° C, mean annual precipitation is 2200 mm (Rollenbeck et al. 2007). A moderate rainy season typically extends from March – July. On average, one month with <100 mm of rain occurs during the driest part of the year, from October – February (R. Rollenbeck, pers. comm.). Shorter dry spells of 1–2 weeks, typically induced by westerly foehn winds, occur more frequently (Emck 2007). Fog is uncommon at this elevation (Rollenbeck et al. 2006, in press).

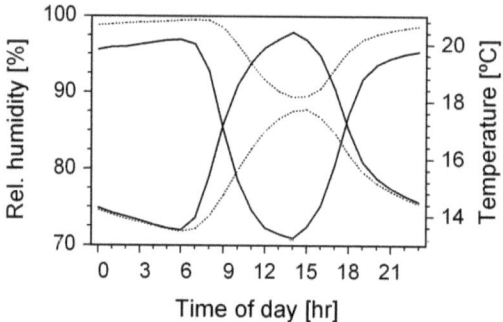

Figure 1 Daily course of relative air humidity and temperature at 2 m height under the crowns of ITs (solid lines) and forest trees (stippled lines). Data from 9 pairs of loggers sequentially set over the course of ca. 1 year.

Data collection

During December 2003 – January 2004, we removed all vascular epiphytes from the trunk base of 93 individuals of *Piptocoma discolor* (Asteraeae), 48 isolated in pastures and 45 in adjacent enclosed forest. *P. discolor* is locally common on slopes and in ravines of both forest and cleared pasture land. The species is characterized by fast growth (Homeier 2004) and its fissured, spongy bark apparently promotes the growth of epiphytic bryophytes and vascular plants. Root bases rarely extend beyond 0.2 m in height. All ITs had established in pastures, as evident from their architecture (short trunks, divaricated crowns) and hence were not remnant trees.

Epiphytes were removed from cylindrical plots 0.5–2.25 m in trunk height, minimizing substrate damage. Total plot area was 84.4 (ITs) and 90.4 m^2 (forest), respectively. Trees were revisited after 3–6 months to remove rare resprouting fragments of creeping plants that were overlooked during initial removal. After 2 years, all vascular epiphyte seedlings colonizing these plots were recorded and identified. Accidental epiphytes, fern gametophytes and plants establishing in knotholes were omitted. Seedlings on ITs were marked with coated steel nails. Plant survival was recorded (again) after one year.

Identifications were based on years-long local field experience. Young sporophytes of Vittariaceae and Dryopteridaceae (*Elaphoglossum*) species are easily confounded and we may have slightly overestimated the former. Young seedlings of the closely related bromeliad genera *Tillandsia* and *Vriesea* are indistinguishable. Because they further share similar ecological requirements, we made no attempt to separate them (combined as

'*Tillandsia*' in the following). For the same reasons, *Pecluma* and *Serpocaulon* (Polypodiaceae) were combined as '*Pecluma*'.

For each plot we recorded percentage cover of lichens and bryophytes, DBH, canopy openness, and, for IT plots, distance to enclosed forest. Canopy openness was measured with a spherical densitometer (Lemmon 1957), distance to forest by means of ArcGIS 9 (ESRI, Redlands, CA, U.S.A.) and a geo-referenced aerial photograph. We logged air temperature and relative humidity at 2 m of height by sequentially running data loggers (Onset Hobo Pro, Pocasset, MA, U.S.A.) on pairs of ITs and forest trees ($n = 9$) during January 2004 – February 2005.

Data analysis

Because the number of taxonomic units increases nonlinearly with area, we downsized all plots to 1 m² prior to the analysis of taxa density (the number of taxa per plot). We did this by taking into account those plants growing on the lowest and highest 0.5 m² of each plot cylinder, in order to avoid bias of the resulting sub-samples from skewed vertical distribution patterns.

Analysis of taxa density at the genus level allowed us to classify plants as either hygrophilous or xerophilous (Werner et al. 2005; F. Werner unpubl. data). At family level, several taxa are heterogenous in this respect (Table 1). We applied individual-based rarefaction ($n = 50$) with 10,000 iterations to compare total species richness between treatments (Gotelli & Entsminger 2006).

Because parametric assumptions could not be matched, we analyzed between-group differences of continuous variables through resampling using PC-Ord 4.25 (McCune & Mefford 1999). We used multi-response permutation procedure (MRPP; Mielke et al. 1982), one-factorial and on squared euclidean distance, applying a weighing factor ($C = n_i - 1/\Sigma(n_i-1)$) which results in a MRPP statistic equivalent to a 2-sample t-test or one-way ANOVA F-test (Mielke et al. 1982). A, the chance-corrected within-group agreement, is a sample size-independent measure of 'effect-size'. When all items are identical within groups, then the observed $A = 1$ is the highest possible value for A. If heterogeneity within groups equals expectation by chance, then $A = 0$. If there is less agreement within groups than expected by chance, then $A < 0$.

Table 1. Epiphyte seedling densities (individuals/m²) on forest trees vs. ITs.

	Forest Mean	Forest SD	ITs Mean	ITs SD	A	$p <$	Ecol. Req.[a]	Disp. Mode[b]
Araceae	0.08	± 0.23	–	± –	0.055	**0.01**	H	A
Anthurium	0.07	± 0.20	–	± –	0.043	**0.05**	H	A
Philodendron	0.02	± 0.13	–	± –	0.001	0.5	H	A
Aspleniaceae (*Asplenium*)	1.90	± 3.71	–	+ –	0.112	**0.0001**	H	W
Blechnaceae (*Blechnum*)	0.01	0.03	–	± –	0.001	0.5	H	W
Bromeliaeae	0.03	± 0.14	0.22	± 0.45	0.061	0.01	H	W
Guzmania	0.01	± 0.05	–	± –	0.001	0.5	H	W
Tillandsia[c]	0.03	± 0.14	0.22	± 0.45	0.067	**0.005**	X	W
Cyclanthaceae (indet.)	0.02	± 0.11	–	± –	0.001	0.5	H	A
Dryopteridaceae (*Elaphoglossum*)	0.36	± 0.73	0.06	± 0.24	0.065	**0.005**	H-HL	W
Ericaceae (indet.)	0.01	± 0.08	–	± –	0.001	0.5	HL	A
Grammitidaceae	0.06	± 0.19	0.10	± 0.27	-0.005	0.5	H-HL	W
Melpomene	0.03	± 0.14	0.10	± 0.27	0.014	0.5	HL	W
Micropolypodium	0.02	± 0.11	–	± –	0.001	0.5	H	W
Alansmia	0.01	± 0.09	–	± –	0.001	0.5	H	W
Hymenophyllaceae	0.31	± 0.49	–	± –	0.169	**0.0001**	H	W
Hymenophyllum	0.16	± 0.35	–	± –	0.086	**0.001**	H	W
Trichomanes	0.16	± 0.38	–	± –	0.074	**0.005**	H	W
Orchidaceae	0.04	± 0.15	–	± –	0.019	0.1	H-X	W
Dichaea	0.03	± 0.14	–	± –	0.012	0.5	H	W
indet.	0.01	± 0.03	–	± –	0.001	0.5	n.a.	W
Piperacae (*Peperomia*)	0.29	± 0.71	–	± –	0.072	**0.0001**	H	A
Polypodiaceae	0.69	± 1.44	0.14	± 0.34	0.057	**0.005**	H-X	W
Campyloneurum	0.01	± 0.09	–	± –	0.001	0.5	H-HL	W
Pecluma[d]	0.65	± 1.43	–	± –	0.090	**0.0001**	H	W
Pleopeltis	0.02	± 0.13	0.14	± 0.34	0.040	**0.05**	X	W
Urticaceae (*Pilea*)	0.02	± 0.12	–	± –	0.012	0.5	H	G
Vittariaceae	2.39	± 5.51	–	± –	0.080	**0.0005**	H	W
Polytaenia	1.93	± 5.04	–	± –	0.062	**0.0005**	H	W
Radiovittaria	0.06	± 0.27	–	± –	0.020	**0.05**	H	W
Vittaria	0.39	± 2.47	–	± –	0.002	0.1	H	W

p-values that remain significant after FDR correction at $p < 0.05$ are in bold.

[a] ecological requirements: H = hygrophilous and shade-loving ; HL = hygrophilous and moderately light-demanding; X = xerophilous and light-demanding. Based on Werner et al. 2005; F. Werner, unpubl. data.

[b] dispersal modes: W = wind-dispersed; A = animal-dispersed; G = gravity-dispersed.

[c] may include *Vriesea* spp.

[d] may include *Serpocaulon* spp.

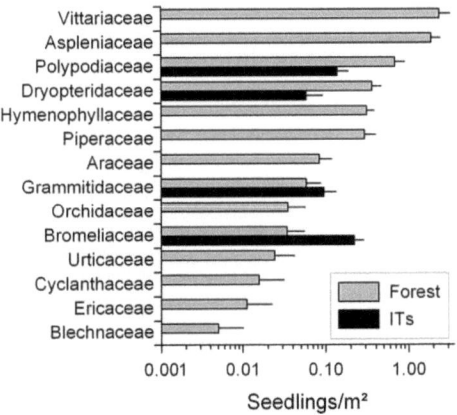

Figure 2 Density of epiphyte seedlings after 24 months of colonization (means and standard errors). Note the logarithmic scale of the graph.

Seedling densities on ITs were zero-truncated, so we used incidence data of 1 m² sub-plots to test for effects of canopy openness, distance to forest and bryophyte cover on epiphyte establishment. Data were analyzed by means of a randomization test (10,000 iterations), after separating IT plots into two equal-sized groups (higher and lower values of canopy openness, distance and bryophyte cover, respectively). Seedling mortality on ITs was analyzed by means of the same randomization procedure for binary data. Due to a low number of observations, we grouped seedling genera as either commonly (> 5% of all epiphyte individuals) or uncommonly (< 0.5%) found as post-juveniles on local ITs (Werner et al. 2005) to test if seedling survivorship is coupled with abundance of post-juveniles.

Where appropriate, multiple tests of significance were corrected for a table-wide false discovery rate (FDR) of $p < 0.05$ according to Benjamini & Hochberg (1995).

Table 2. Plot characteristics.

	Forest		ITs			
	Mean	SD	Mean	SD	A	p
Host DBH [cm]	37.88 ±	13.49	34.18 ±	12.34	0.010	0.172
Plot size [m²]	2.01 ±	0.74	1.76 ±	0.60	0.023	0.074
Canopy openness [%]	17.03 ±	4.47	68.08 ±	15.65	0.828	< 0.0001
Distance to forest [m]	– ±	–	189 ±	117	–	–
Lichen cover [%]	9.03 ±	8.32	47.50 ±	15.45	0.703	< 0.0001
Bryophyte cover [%]	38.94 ±	20.63	17.55 ±	13.40	0.273	< 0.0001

RESULTS

Mean temperature was 16.4° C ± 0.8 SD (ITs) and 15.3° C ± 0.6 (forest), mean relative humidity was 87.6% ± 2.1 and 96.3% ± 2.8 respectively. Hourly means differed significantly ($p < 0.05$; Wilcoxon test) from 7:00–18:00 hr (T_{mean}), and for the entire day (RH_{mean}), respectively. Differences in hourly means peaked at 10:00–11:00 a.m. (3.3° C) and 11:00–12:00 a.m. (19.8 %; Fig. 1). Lichen cover was significantly higher on the stem bases of ITs vs. forest trees ($A = 0.703$, $p < 0.0001$), whereas bryophyte cover was lower ($A = 0.273$, $p < 0.0001$; Table 2).

Overall, the IT plots yielded 48 seedlings from four genera and families, the forest plots 533 seedlings from 24 genera and 13 families (Fig. 2; Table 1). Although forest seedlings were composed of anemochorous (91%), zoochorous (8%) and barochorous taxa (1%), seedlings on ITs comprised exclusively anemochorous taxa. The underrepresentation of zoochorous relative to anemochorous taxa on ITs was significant ($p < 0.0001$; randomization test).

Density of seedlings was significantly smaller on ITs relative to forest trees ($A = 0.238$, $p < 0.0001$), averaging 0.51 seedlings per m² ± 0.72 SD compared to 6.21 ± 7.21 on forest plots. The number of both families and genera encountered on 1 m² sub-plots was also smaller, each measuring 0.40 ± 0.64 on ITs, whereas forest sub-plots harboured 1.69 ± 1.35 families and 1.73 ± 1.42 genera per m², respectively. These differences were highly significant ($A = 0.274$, $p < 0.0001$ and $A = 0.269$, $p < 0.0001$, respectively).

Total richness in epiphyte families and genera was significantly lower on ITs. Richness of the forest sample rarefied to the size of the IT sample (48 individuals) was 7.92 ± 1.22 families (95% conf. interval = 6–10 families) and 10.35 ± 1.54 genera (95% conf. interval = 8–14 genera) (Fig. 3).

ITs further differed substantially in seedling composition from forest trees. The xerophilous genera *Pleopeltis* and *Tillandsia* were significantly more abundant on IT plots than on forest plots, whereas numerous hygrophilous taxa were significantly less abundant (Table 1). For instance, the fern families Aspleniaceae and Vittariaceae which dominated forest plots were entirely absent from ITs.

Seedling density on IT plots was negatively related to distance to forest and canopy openness, and positively related to bryophyte cover (Fig. 4). The likelihood of colonization (1 m² sub-plots) differed significantly regarding distance to forest ($p = 0.026$), but not for canopy openness or bryophyte cover ($p = 0.086$ and $p = 0.230$, respectively).

During the third year of study, seedling mortality on ITs averaged 25.0%. Seedlings of genera that are commonly found in post-seedling stages on ITs suffered significantly lower mortality than seedlings of genera that are uncommon ($p = 0.040$). Among the former genera (*Pleopeltis* and *Tillandsia*) only 18% of plants died, whereas the latter genera (*Elaphoglossum* and *Melpomene*) exhibited 43% of mortality (Table 3).

Table 3. Establishment and mortality of seedlings, and the representation of their respective (sub-) adult stages on ITs.

		Seedlings	Post-juveniles [a]	
			Rel. abundance on ITs	Abundance ratio
Genus	N	Mortality [%/a]	[% of ind.]	ITs/forest trees
Elaphoglossum	5	60.0	0.13	0.0002
Melpomene	9	33.0	0.25	0.0001
Pleopeltis [b]	15	6.67	7.0 [b]	1.10
Tillandsia [c]	19	26.3	47	0.34

[a] calculated from Werner et al. (2005) after exclusion of one outlier species (*Dryadella werneri*).
[b] note: abundant creepers tend to be underestimated by the 'stand' concept employed by Werner et al. (2005).
[c] may include *Vriesea* spp.

DISCUSSION

Post-juvenile assemblages of vascular epiphytes on remnant trees at our site are substantially less abundant (by 85%) and diverse (80% of species per tree) 10–30 years after their isolation in pastures (Werner et al. 2005). This impoverishment affects mesophilous and hygrophilous species, whereas xerophilous canopy taxa remain relatively well-represented. This compositional skew of post-juvenile assemblages was mirrored by the patterns of seedling establishment we observed on ITs after two years. Seedlings of mesophilous and hygrophilous taxa were restricted to few individuals in the genera *Elaphoglossum* and *Melpomene*, whereas the great majority of seedlings was from the xerophilous genera *Pleopeltis* and *Tillandsia* (Fig. 2).

Seedlings of animal-dispersed taxa were relatively scarce even in forest plots, reflecting the domination of the local epiphyte flora by wind-dispersed taxa (Homeier & Werner, in press; Lehnert et al., in press). Moreover, endozoochorous seeds have a low probability for attaching to near-vertical surfaces such as trunks. The absence of animal-dispersed taxa from ITs intuitively suggests stronger dispersal constraints in zoochory vs. anemochory. However, since local animal-dispersed taxa share relatively high humidity requirements (Werner et al. 2005), increased desiccation stress on ITs (Fig. 1) may cause a similar pattern.

The rain of wind-dispersed diaspores should decrease geometrically with growing distance from the source (Madison 1979). Our data revealed a significant effect of distance to forest on seedling establishment, which shows that diaspore rain was reduced on ITs. However, the effect of distances to forest on IT seedling densities was rather small (Fig. 4). Forest vegetation is not necessarily the sole source of diaspores for IT assemblages, as reproductive adults especially in the genera *Pleopeltis* and *Tillandsia* also occur on ITs (Table 3; Werner et al. 2005). By adding to the diaspore rain from forest sources, such plants may dilute the effects of dispersal constraints. However, while diaspores originating from ITs may have weakened the relationship between total seedling density and distance to forest, they cannot help explain the scarcity of hygrophilous taxa on ITs. Because the studied ITs formed regrowth, their entire epiphytic flora (including reproducing adults) must have established in isolation, subjected to the same establishment constraints as seedlings in our study.

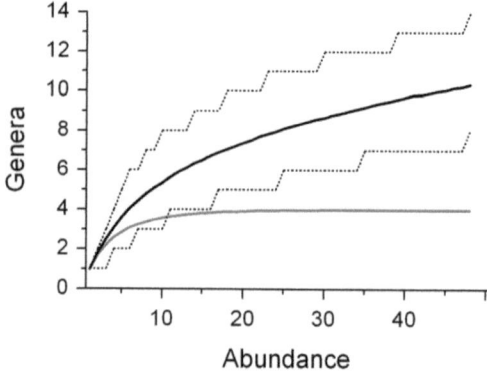

Figure 3. Individual-based rarefaction (10,000 iterations) of generic richness in the forest (black line) and on ITs (grey line). Dotted lines reflect the respective 95% confidence intervals for the forest sample as determined from the 0.025 and 0.975 frequency values in the simulated data.

In our study, hygrophilous understorey specialists characterized seedling assemblages of forest plots (e.g., *Asplenium*, *Pecluma*, Vittariaceae spp.) but were entirely absent from ITs (Table 2). Instead, IT seedling assemblages were strongly predominated by xerophilous taxa that were poorly represented in the forest understorey, despite of their common occurrence in the forest canopy. A corresponding paucity in understorey taxa has been reported from ITs and disturbed forests elsewhere (Barthlott et al. 2001; Krömer & Gradstein 2003; Flores-Palacios & García-Franco 2004; Hietz 2005), and cannot be explained easily by other factors than microclimatic changes (Fig. 1).

ITs can be viewed as forest fragments that are exposed to multiple edge effects, including increased light levels, wind velocity, temperature and reduced air humidity (Laurance 2004). The resulting harsher microclimate strongly affects many organisms, including epiphytic lichens and bryophytes (Moen & Jonsson 2003; Hylander 2005). In our study, however, IT plots exposed to higher light levels (greater canopy openness) did not show a significantly lower probability of colonization. This may be related to the great stochasticity inherent to establishment dynamics throughout (Hubbell 2001; Laskurain et al. 2004), coupled with an unexpectedly low number of observations (only 48 seedlings in 21 of 48 plots). Moreover, canopy openness is not an ideal measure of exposure. Although canopy openness is a good proxy for wind penetration, air humidity and temperature in forest understorey, these parameters are presumably little influenced by canopy properties of solitary trees. Canopy openness thus reflects only one of several major components of physical edge effects that affect water budgets in plants. The measured differences in canopy openness between IT plots may further be of negligible relevance for epiphytes that already perceive even the most shaded IT plots as exceedingly exposed.

Light levels, temperature and relative humidity were greatly altered around IT trunk bases (Fig. 1; Table 2), to the point of closely resembling conditions in the upper forest canopy (F. Werner and C. Gehrig, unpubl. data). Upper canopy conditions on IT trunk bases were mirrored by patterns of non-vascular epiphytes. Lichens, most of which are sensitive to excessive humidity, flourished, whereas bryophytes, which favour constant humidity (Nöske et al. in press), exhibited low covers (Table 2). Regardless of diaspore influxes, it seems most unlikely that seedlings of vascular understorey epiphytes may establish successfully in such a harsh, canopy-like environment. Hietz & Briones (1998, 2001) showed that vertical stratification of (adult) epiphytic ferns closely reflects exposure tolerance, being correlated with a wealth of morphological and physiological traits that influence rates of uncontrolled water loss. Zotz and co-workers could further demonstrate that water relations are strongly influenced by the surface-volume ratio and hence plant size in vascular epiphytes (Zotz et al. 2001; Zotz & Hietz 2001).

Mortality rates on ITs were remarkably low (25% on average), despite an unusually dry year of 2006 (R. Rollenbeck, pers. comm.). Much higher mortality rates – particularly due to drought – have been found in early epiphyte seedlings elsewhere (Benzing 1978; Larson 1992; Laman 1995; Tremblay 1997; Zotz 1998; Castro Hernández et al. 1999; Hietz et al. 2002; Zotz et al. 2005). Moreover, seedling mortality declines drastically with age (Castro Hernández et al. 1999; Mondragón et al. 1999; Hietz et al. 2002; Zotz et al. 2005), suggesting that many of the seedlings recorded after 2 years had already passed an intense phase of selection.

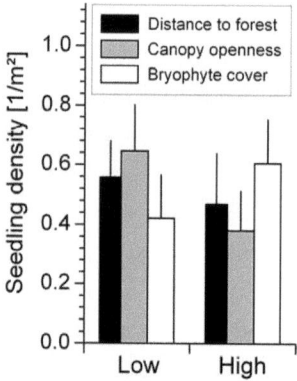

Figure 4. Effects of the predictor variables distance to forest, canopy openness and bryophyte cover on seedling densities on ITs (means and standard errors). Bars reflect the seedling density on each 24 plots of highest or lowest predictor values respectively, with ranges and means as follows: distance to forest low: 10–129 m (mean 92), high: 132–432 (287); canopy openness low: 22–72 (57), high: 72–92% (79); bryophyte cover low: 3–13% (7.4), high: 15–58 (28). Note that the shown data was standardised to 1 m² plot size and binarised prior to significance testing.

Diverging survival rates add further support to the notion that seedling assemblages were shaped by mortality. The genera *Pleopeltis* and *Tillandsia* exhibited lower mortality than the genera *Melpomene* and *Elaphoglossum*. Adults of the former genera locally abound on ITs, where adults of the latter genera are rarely found (Table 3), suggesting that seedling mortality differs between epiphyte taxa according to their predisposition for life under high levels of exposure. This conclusion implicates that diverging rates of seedling mortality (filtering) shape the composition of post-juvenile assemblages. The fact that post-juvenile stages of *Melpomene* and *Elaphoglossum* favour growth sites with higher moisture levels than *Pleopeltis* and *Tillandsia* further suggests that the tolerance of seedlings to drought is a major predictor of seedling mortality. For instance, *Pleopeltis macrocarpa*, a poikilohydrous fern that regionally extends into perarid interandean forest (Werner & Gradstein, in press) accounted for almost one third of all seedlings recorded on ITs and showed the highest survival rate at our site (Table 3). Thus, our results strongly suggest that both decreased diaspore rain and drought-related seedling mortality reduce seedling densities on ITs, and that increased physical exposure rather than properties of diaspore rain shape the floristic composition of IT seedling assemblages.

The study of establishment limitations is a key for the understanding of current and future patterns of epiphyte diversity. Drought inflicts a major challenge for adults of many epiphyte species (e.g., Zotz & Tyree 1996; Benzing 1998), and seedlings are even more drought-sensitive (Zotz et al. 2001). Desiccation stress increases with structural forest disturbance and – in most regions – with atmospheric warming (Laurance 2004; Malhi & Phillips 2004). Thus, abiotic seedling requirements may increasingly constitute a bottleneck for the persistence of epiphyte populations.

Many if not most vascular epiphyte species require 1–2 decades to reach maturity (Larson 1992; Benzing 1998; Zotz 1998; Hietz et al. 2002; Schmidt & Zotz 2002). Since many of these species cannot be identified much earlier, classic ecological inventories will invariably record consequences of changing land use and climate only with considerable time lag. Even experimental approaches with post-seedling life stages (e.g., Nadkarni & Solano 2002) can only provide limited insight in this regard, as they do not take into account seedling requirements and may therefore underestimate the sensitivity of species greatly. Thus, a better understanding of seedling requirements and performance is needed to construct meaningful climate envelopes, and predict the development of epiphyte communities after habitat modification and with global climate change.

Conclusions

With the exception of few hardy canopy taxa, we found that rates of establishment of vascular epiphytes on ITs were much smaller than on corresponding forest trees. Thus, altered recruitment is apt to explain the dramatic and directional impoverishment of post-juvenile epiphyte assemblages on remnant trees, which is locally observed after prolonged isolation in pastures. Although isolated trees in anthropogenic landscapes are key structures for the maintenance of forest biodiversity in many aspects (Manning et al. 2006), our results suggest that their value for the conservation of epiphytes can be very limited.

Though we found evidence that seedling establishment on ITs was constrained by dispersal, the results imply additional establishment limitations. Patterns of floristic seedling composition and mortality suggest a major influence of increased desiccation stress on IT seedling assemblages, which may greatly exceed the influence of dispersal constraints (compare Snäll et al. 2003). However, we could not test this notion directly, due to limitations in our study design and the small number of seedlings. Disentangling the roles of these diversity drivers in complex anthropogenic landscapes poses methodological difficulties that may have led to the frequent overestimation of the role of dispersal limitation in epiphytes (Pharo & Zartman 2007; Werth et al. 2007). Given that establishment constraints are key for the prediction of future epiphyte communities, further studies concerning this matter will yield critical insights.

Chapter 7

INCREASED MORTALITY OF VASCULAR EPIPHYTES ON ISOLATED TREES FOLLOWING FOREST CLEARANCE IN MOIST MONTANE SOUTH ECUADOR

ABSTRACT

Epiphytes in fragmented and degraded tropical forests may suffer from the effects of increased exposure to atmospheric conditions as they prevail above forest canopies. We studied the response of a vascular epiphyte assemblage to severe structural forest disturbance in a montane moist forest in Ecuador. Well-established individual plants were recorded on isolated remnant trees (IRTs) in a fresh clear-cut and in undisturbed forest (controls), and their growth and survival was followed during three consecutive years. Wind-throw and branch breakage accounted for the loss of 24% of plants on IRTs but for < 1% in the forest. Plant losses from intact phorophyte structures were also higher on remnant trees than on forest trees, totalling 72% vs. 11% in 3 years. Losses on IRTs were greatest after the first year (52%) and among ferns and dicots, and lowest after the second year (20%) and among monocots (aroids, bromeliads, orchids). Plants that remained alive on IRTs commonly showed a marked decrease in maximum leaf length. Annual increment in leaf number varied more widely, both between and within epiphyte taxa. The present study provides first experimental, field-based evidence that increased physical exposure affects the performance of well-established vascular epiphytes. Our results suggest that growth conditions may often be a more influential driver of vascular epiphyte diversity in disturbed habitats than dispersal constraints.

Key words: bryophytes, diversity, edge effects, forest fragmentation, growth, human disturbance, microclimate, species richness, scattered trees, tropical montane forest

INTRODUCTION

While tropical deforestation proceeds at high pace, conservation biologists remain divided over the extent to which anthropogenic habitats will be able to offset the loss of biodiversity from primary forests (Putz et al. 2001; Laurance 2006; Gardner et al. 2007). Globally, tropical secondary forests have reclaimed a sixth of the primary forests cleared in the 1990s and may soon exceed primary forest habitats in extent (Brown & Lugo 1990). Most of the area cleared of primary forest, however, is permanently or repeatedly taken into agricultural or pastoral use, creating extensive anthropogenic landscapes. These landscapes typically are complex mosaics of agricultural matrix habitat and regrowth isolating fragments of primary vegetation. Currently, primary vegetation is already highly fragmented in most tropical areas, and fragment size tends to decline fast, dividing populations and creating degraded edge habitats (Gascon et al. 1999). A new dimension of the fragmentation dilemma has been entered with the onset of global climate change. Even narrow matrix bands threaten biodiversity by hindering compensatory migration, such as upslope migration across the deforested foothills of the northern Andes (Bush 2002). It has thus become increasingly apparent that understanding how species are affected by habitat fragmentation requires information on their responses to all components of the landscape, including secondary forest and matrix habitats (Gascon et al. 1999).

Vascular epiphytes are a major element of tropical forest structure, functionality, and biodiversity. Various recent studies have reported impoverished and compositionally biased epiphyte communities from disturbed habitats. However, the processes governing these changes remain poorly resolved. Virtually all available studies dealing with epiphytes in disturbed habitats are descriptive and base on chrono-sequences; while this 'space-by-time substitution' approach is powerful at revealing patterns, it is limited at resolving underlying processes (cf. Laube & Zotz 2006).

At present, one of the questions most widely debated and pursued among students of human disturbance effects on epiphytes is the importance of growth conditions ('local factors') vs. dispersal limitations ('regional factors') for the persistence of epiphyte populations (Pharo & Zartman 2007). This debate extends across the major epiphyte taxa, and while some authors contend the view that shifted abiotic conditions are of paramount importance, others stress the influence of dispersal constraints. The fact that epiphytes (including lichens and bryophytes) tend to colonize phorophytes non-randomly, aggregated around established conspecifics, has often been interpreted as indirect evidence for dispersal constraints (e.g., Benzing 1981; Vandunné 2002; Cascante-Marin et al. 2006; Laube 2006), although other factors (microclimate, substrate suitability, mycorrhizal or animal mutualisms) are apt to produce equivalent patterns (Johansson 1974; Madison 1979; Yeaton & Gladstone 1982; Tremblay et al. 1998; Krömer & Gradstein 2003; Lehnert 2007). Indeed, recent work suggests that the importance of dispersal constraints has been overestimated, at least for non-vascular epiphytes (Pharo & Zartman 2007; Werth et al. 2006), which are much better studied in this respect than vascular epiphytes.

In montane moist SE-Ecuador, Werner et al. (2005) studied vascular epiphytes on isolated remnant trees (IRTs) 10–30 yr after their isolation in pastures. They found epiphyte assemblages on IRTs greatly impoverished relative to adjacent closed forest, with the number of individuals and species per tree reduced by 80%, and estimated total species richness reduced by 71% (Werner et al. 2005; Nöske et al. 2008). The authors speculated that such pronounced impoverishment may be caused by a combination of increased mortality in well-established plants (post-seedling life-stages) coupled with reduced recruitment. In the present paper we test the hypothesis that increased mortality of well-established epiphytes can explain the long-term impoverishment of epiphyte assemblages observed by Werner et al. (2005). We do so by following the fate of well-established epiphytes through the first 3 years after the isolation of their phorophytes in a fresh clear-cut.

METHODS

Study site

Field work was carried out from 2003–2006, at 2000 m elevation in the surroundings of Estación Científica San Francisco (ECSF) near Podocarpus National Park in Zamora-Chinchipe Province, southeast Ecuador (3°58' S, 79°04' W). Potential vegetation of slopes and ravines is moist forest with a canopy height of ca. 15–20 m (Homeier et al., in press). The area fosters a remarkably rich epiphytic and terrestrial flora, regarding both total species numbers and the representation of endemics (Brehm et al., in press; Homeier & Werner, in press; Lehnert et al., in press; Richter, in press). Mean annual temperature is 15.5 °C, mean annual precipitation is 2200 mm (Rollenbeck et al. 2007); fog is uncommon at this elevation (Rollenbeck et al., in press). A gentle rainy season typically extends from April – July. Since the beginning of climate recording in 1998, on average a single month with <100 mm has occurred during the driest part of the year from September – February (R. Rollenbeck, pers. comm.). Shorter dry spells of 1–2 weeks, typically induced by westerly winds (foehn), occur more frequently (Emck 2007). Regular climatic conditions prevailed during 2003–2005, but 2006 was unusually dry (M. Richter, unpubl. data).

Sampling

Sampling was initiated 3–4 days after clearance of a ca. 2 ha large old-growth forest patch by a local colonist in September 2003 (Fig. 1). The wedge-shaped clearing bordered old-growth forest to one side and a ca. 5 yr old pasture to the other; the third, considerably shorter side of the clearing bordered a second pasture of ca. 20 yr age. The clearing faced NNW, slope inclination was 25–30°, original canopy height had been ca. 15 m. IRTs were of over 30 species, apparently conserved with little or no bias regarding their taxonomic identity.

We sampled all facultative and obligate vascular epiphytes accessible with a steel ladder (up to 5 m height) on IRTs, only omitting phorophytes that were dead, visibly damaged during forest clearance, or at < 20 m distance from the former edge. Early juvenile stages were omitted for their differing (higher) rates of mortality and relative growth (Larson 1992; Zotz 2004b; Zotz et al. 2005; Winkler et al. 2005, 2007), and the difficulties in marking such individuals unequivocally without provoking their damage or loss. Individual plants were marked with coloured multi-fibre nylon strings of 1–2 mm in diameter and (in subsequent years) numbered aluminium tags and -wire, carefully avoiding damage or hazard to the plants. In filmy ferns (Hymenophyllaceae), it was frequently not possible to define individual plants unambiguously, and the sampled clusters may therefore represent more than one individual. In bromeliads, ramets were counted as parts of one plant unless they were fully separated. All plants were photographed, identified, and their position on the phorophytes was recorded (height, cardinal direction). Leaf number and length of the largest leaf (including the stem in pleurothallidinid orchids) were determined during each sampling campaign, except for plants where these measures could not be taken accurately or unequivocally (especially filmy ferns).

In the course of the study, bunches of a forage grass (*Setaria sphaceolata*) were planted and terrestrial regrowth carefully controlled mechanically at an annual basis, but fire was not used. Control trees were chosen randomly in mature forest adjacent to the clearing at a minimum distance of 40 m from the edge, and along two trail sections at corresponding altitude and cardinal exposition within the Reserva Biológica San Francisco, located 2–2.5 km from the clearing under study and in the same valley. Sampling was repeated after 12, 24 and 36 months (September – October of 2004, 2005 and 2006).

We distinguished two main types of plant loss: (1) dislodgement with their woody substrate ('substrate failure'; phorophyte uprooting or snapping, branch fall), and 2) plant loss unrelated to the fate of phorophytes. In the latter case, which was the main focus of our study, we further distinguished between a) deceased in situ (remains of shoot still present), and b) disappeared. For convenience, we will refer to the loss of epiphytes from intact phorophyte structures as mortality. Actual survivorship was assessed in 2006 for plants affected by substrate failure, but not for plants classified as disappeared, due to the difficulties encountered in locating such plants among debris.

The percentage surface cover of living bryophytes (at 0.25–5 m height on trunks) was estimated for all intact phorophytes after three years (in 2006). Hemispherical photographs were taken at 2 m height under the crowns of nine pairs of trees (IRTs, control) using a NIKON digital camera and 180°- fish-eye lens.

Analysis

We treated the individual plant as the independent sample unit. Since early juveniles were omitted, we presumed that all studied plant individuals were well-adapted to their specific growth site, and responded primarily to treatment effects, regardless of the respective host tree. This assumption was corroborated by a mortality analysis at phorophyte level that showed low variability within treatments and highly significant differences between treatments (see results section). In fact, variability within treatments was unexpectedly low, considering that differences in epiphyte composition between individual phorophytes must have added considerable noise to mortality rates at phorophyte level. Our analyses focus on the family level since patterns at genus level are congruent (Appendix 5).

Hemispherical photographs were analysed using the software GLA 2.0 (SFU, Burnaby, BC, Canada). Calculations of light transmission are based on the mean daily transmission value of 3,731 Whm^{-2} measured at the study site (Emck 2007).

Figure 1. The clearing under study few days after clear-cutting (Sept. 2003), with the newly created forest edge on the left margin of the photograph (a); (b–e): responses of epiphytes to exposure on isolated trees: (b) *Lellingeria subsessilis* (Grammitidaceae) on IRT during the first census (Sept. 2003); (c) the same plants died back after 1 yr, with both leaf no. and max. leaf length substantially reduced; note also the decay of bryophytes; (d) *Elaphoglossum guamanianum* (Dryopteridaceae) after 1 yr (Sept. 2004) with 13 leaves up to 19.7 cm length; this fern had possessed 5 large leaves up to 39.0 cm length at the previous census and 19 leaves up to 9.6 cm at the following census; (e) *Guzmania killipiana* (Bromeliaceae) after 1 yr; note the desiccated leaf apexes that had reduced max. leaf length by 39% within 1yr (leaf no. had not changed).

Binary data were analysed by means of Fisher's exact test (two-sided) except for comparisons of epiphyte losses between treatments, where we used a randomisation test (one-sided, 10,000 permutations). Continuous data were analysed by means of the Mann-Whitney U-test unless indicated otherwise, using Statistica 6.0 (Tulsa, OK, U.S.A.).Analyses of mortality unrelated to phorophyte fate were done after exclusion of the respective phorophytes (wind-throw) or individual epiphytes affected (branch-breakage).

RESULTS

As expected, mean daily light transmission (direct and diffuse) was substantially higher below IRT crowns ($p < 0.0001$), averaging 2.27 kWhm^{-2} ± 0.45 (mean ±SD) on IRTs vs. 0.43 kWhm^{-2} ± 0.19 (forest). Three years after clear-cutting, the cover of living bryophytes on IRTs was 8.8% ± 1.8, substantially lower than on forest trees (57.1% ± 12.4; $p < 0.0001$).

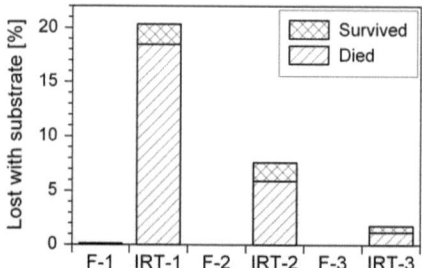

Figure 2. Epiphyte losses due to phorophyte failure (wind-throw, branch breakage) in undisturbed forest (F) and from remnant trees (IRT) over the course of three years following forest clearance (1–3), including the share of these plants that were alive by the time of the last census in 2006.

Mortality due to phorophyte failure

Epiphytes on IRTs had a significantly higher probability of dislodgement with their woody substrate over the three study years ($p < 0.0001$); 23.9% of IRT plants fell but only 0.1% of control plants. Eight out of 45 IRT phorophytes were uprooted or snapped during the first study year, and one during the second. A single IRT branch-fall occurred during the third year. In contrast, wind-throw did not affect any of the 65 forest phorophytes, and branch-

Table 1. Plant losses (mortality [%]) unrelated to phorophyte failure (wind-throw, branch breakage) over the course of three years following forest clearance.

Taxa	2003-2004 Forest		2003-2004 IRTs			2004-2005 Forest		2004-2005 IRTs			2005-2006 Forest		2005-2006 IRTs			2003-2006 Forest	2003-2006 IRTs
	n	Mort.	n	Mort.	P	n	Mort.	n	Mort.	P	n	Mort.	n	Mort.	P	Mort.	Mort.
Araceae	22	0.0	13	15.4	>0.1	22	0.0	11	0.0	1	22	9.1	10	20.0	>0.2	9.1	32.3
Aspleniaceae	43	4.7	27	92.6	<0.0001	41	4.9	2	50.0	n.a.	39	17.9	1	0.0	n.a.	25.6	96.3
Bromeliaceae	232	2.6	99	29.3	<0.0001	224	3.1	70	12.9	<0.005	216	5.1	58	27.6	<0.0001	10.4	55.4
Dryopteridaceae	113	1.8	39	61.5	<0.0001	111	1.8	15	46.7	<0.0001	109	1.8	7	0.0	1	5.3	79.5
Grammitidaceae	36	0.0	27	51.9	<0.0001	36	5.6	13	38.5	<0.01	34	17.6	8	25.0	>0.2	22.2	77.8
Hymenophyllaceae	61	0.0	63	73.0	<0.0001	61	0.0	17	35.3	<0.001	61	3.3	11	45.5	<0.002	3.3	90.5
Orchidaceae	201	3.5	107	41.1	<0.0001	193	2.6	63	14.3	<0.002	187	6.4	53	18.9	<0.01	12.0	59.1
Piperaceae	47	2.1	14	50.0	<0.0001	46	0.0	7	14.3	>0.2	45	11.1	5	60.0	<0.02	13.0	82.9
Polypodiaceae	28	3.6	11	72.7	<0.0001	27	3.7	3	33.3	n.a.	26	7.7	2	0.0	n.a.	14.3	81.8
Vittariaceae	45	0.0	28	64.3	<0.0001	45	2.2	10	40.0	<0.01	44	4.5	5	40.0	>0.05	6.7	87.1
Others	19	0.0	18	72.2	<0.0001	19	0.0	5	20.0	>0.2	19	10.5	4	75.0	<0.01	10.5	94.4
Asteraceae	–	–	1	100.0	–	–	–	–	–	–	–	–	0	–	–	–	100.0
Blechnaceae	1	0.0	1	100.0	–	1	0.0	–	–	–	1	100.0	0	–	–	100.0	100.0
Clusiaceae	1	0.0	4	50.0	–	1	0.0	2	0.0	–	1	0.0	2	100.0	–	0.0	100.0
Cyclanthaceae	–	–	1	100.0	–	–	–	–	–	–	–	–	0	–	–	–	100.0
Davalliaceae	1	0.0	–	–	–	1	0.0	–	–	–	1	0.0	–	–	–	0.0	–
Ericaceae	6	0.0	5	80.0	–	6	0.0	1	0.0	–	6	0.0	1	0.0	–	0.0	80.0
Gesneriaceae	2	0.0	1	0.0	–	2	0.0	–	–	–	2	0.0	1	100.0	–	0.0	100.0
Lycopodiaceae	7	0.0	3	66.7	–	7	0.0	1	100.0	–	7	0.0	0	–	–	0.0	100.0
Urticaceae	1	0.0	2	100.0	–	1	0.0	–	–	–	1	100.0	–	–	–	100.0	100.0
Monocots	455	2.9	220	34.5	<0.0001	439	2.7	144	12.5	<0.0001	425	5.9	121	23.1	<0.0001	11.1	56.0
Dicots	57	1.8	27	59.3	<0.0001	56	0.0	11	9.1	>0.2	55	10.9	9	66.7	<0.0001	12.5	87.7
Pteridophytes	335	1.5	199	69.3	<0.0001	330	2.4	61	41.0	<0.0001	322	6.8	34	26.5	<0.005	10.4	86.7
Total	847	2.2	446	51.6		825	2.4	216	20.4		802	6.6	164	26.2		10.9	71.5

fall only affected a single epiphyte in the course of the study. After 3 years, only 11.4% of 140 dislodged IRT plants were still alive (Fig. 2).

Mortality unrelated to phorophyte fate

Excluding losses due to wind-throw and branch breakage, the mean mortality on phorophytes with 5 or more epiphyte individuals was 69.9% ± 15.0 and 11.5% ± 9.7 for IRTs and forest trees, respectively, over 3 years. Similar results were obtained when including all phorophytes (Fig. 3). These differences were highly significant (both $p < 0.0001$), showing that individual characteristics of phorophytes had only minor influence on the fate of individual plants relative to treatment effects.

Overall, the total mortality of individual plants was 51.6% from IRTs and 2.2% in the forest after the first year following forest clearance. Mortality on IRTs was elevated significantly throughout the entire spectrum of common epiphyte taxa. Rates of mortality on IRTs differed significantly between taxonomic groups (Table 1; Fig. 4). For instance, first year mortality was significantly higher in dicots ($p < 0.02$) and ferns ($p < 0.0001$) than in monocots, reflecting relatively low mortality among the principal monocot families (Araceae, Bromeliaceae, Orchidaceae).

Total mortality on IRTs had declined significantly to 20.4% during the 2nd study year ($p < 0.0001$), but was still significantly increased relative to forest levels (2.4%; $p < 0.0001$). During the 3rd year, mortality on IRTs was not significantly higher than in the previous year (26.2%; $p = 0.21$). Third-year mortality of epiphytes was also higher among control plants (6.6%) than in the previous study year ($p < 0.0001$). Nonetheless, 3rd year mortality on IRTs remained significantly elevated relative to forest levels ($p < 0.0001$). At family level, a 2nd yr decrease in mortality on IRTs was shown by all larger families. For four families, this decrease was significant (Bromeliaceae, Drypteridaceae, Hymenophyllaceae, Orchidaceae; all $p < 0.05$). The following 3rd yr showed a more complex pattern with mortality continuing to decrease in several families of ferns, but increasing in most angiosperm families relative to the previous year (Table 1). However, only in Bromeliaceae did mortality differ significantly ($p < 0.05$) from the previous year.

Of the plants lost from intact IRTs in the course of 3 years, 29.0% were classified as 'disappeared' whereas the remainder plants (69.0%) evidently died in situ. A slightly lower percentage of lost plants (22.8%) were classified as disappeared among control plants (forest).

Growth

Among plants surviving on IRTs, a marked decrease in maximum leaf length was observed in most taxa. Annual increment in leaf number varied more widely, both between and within epiphyte families (Fig. 5). During one of the study years, Dryopteridaceae (2nd yr) and Orchidaceae (3rd yr) even showed significantly increased increment in leaf number on IRTs (Appendix 6).

DISCUSSION

Mortality due to phorophyte failure

Epiphytes on IRTs were more frequently dislodged with their woody substrate than plants on forest trees. This was essentially a consequence of uprooting and snapping of trees; branch breakage had only negligible effects, related to the fact that the great majority of sampled plants grew on trunks rather than on branches. Isolated trees in clearings are subjected to high wind velocity and turbulence (Flesch & Wilson 1999), which can increase wind-throw dozens of meters deep into tropical forest edges (Laurance et al. 1998).

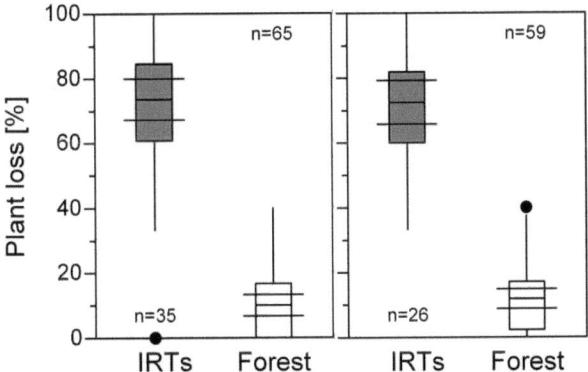

Figure 3. Rates of mortality from all phorophytes (left) and phorophytes with 5 or more plant individuals (right) 3 yr after isolation of trees in a clearing (IRTs) relative to controls (forest). Phorophytes lost in the course of the study are excluded. Outside values (lower quartile plus 1.5 interquartile range) are shown as black dots, notches (95% confidence intervals [median ± 1.58 interquartile range /sqrt n]) are shown as horizontal lines.

IRT epiphytes fallen with their woody substrate had a low likelihood of survival. In Florida woodlands, almost 50% of bromeliad individuals died within 6 months after falling to the ground (Lowman & Linnerooth 1995; see also Benzing 1981; Mondragón et al. 2004). Matelson et al. (1993) documented a leptokurtic decline in the survivorship of fallen epiphytes in a Costa Rican cloud forest, with only 7% of plants surviving the 21-mo study period. Here, mortality did not differ significantly between plant families and was negatively correlated with light levels, suggesting that excessive shade and humidity played a central role. In the case of our study clearing, a similar scenario would be expected from plants buried under logs or coming to rest very near to the ground, where they are quickly outshaded by vigorous terrestrial regrowth. Plants coming to rest on top of logs may be more likely to succumb to excessive exposure. In any case our figures mortality are likely to be unrepresentatively low, given that forest-clearance in the region is usually followed by burning of logs and intensive cattle farming.

Mortality unrelated to phorophyte fate

The dynamic character of the epiphytic habitat has often been emphasised in the literature. With annual mortality rates of 2.2 and 2.4%, epiphyte mortality in the forest (controls) was surprisingly low after the first two years of study. The single study published, to our knowledge, on mortality rates of an entire assemblage of vascular epiphytes reports much higher values of 56% mortality over five years from a lowland rain forest in Panama (Laube & Zotz 2006). However, that study was restricted to one phorophyte species (the palm *Iriartea deltoidea*), on which epiphyte survival may be greatly reduced by falling fronds brushing plants off the trunk (Laube & Zotz 2006). Mortality rates of well-established plants from selected taxa were also somewhat higher in several studies from a montane moist Mexican forest, even when excluding branch breakage (Hietz 1997; Hietz et al. 2002; Winkler et al. 2007). These differences may be related to different study designs. In our study, only under- and lower midstorey epiphytes were sampled, most of which grew on trunks. Such plants should have a considerably lower likelihood of being dislodged than plants on canopy branches, which are subjected to stronger wind turbulence (but see Zotz & Schmidt 2006). But even when excluding the impact of branch fall and dislodgement from substrate, mortality may generally be lower for epiphytes in the understorey than in the canopy, since life in the understorey certainly constitutes a 'compromise' for epiphytes not only in terms of opportunities (light availability) but also risks (desiccation). Thus, our results on mortality and growth rates are in line with experimental studies on secondary succession of disturbed inner crowns that revealed a very slow development of epiphyte assemblages (Nadkarni 2000; Cobb et al. 2001). In any case, the considerable between-year variability in mortality (Fig. 4) suggests that three years of observation is too short to permit a general interpretation of trends, given that many vascular epiphytes have life-cycles in the range of 1–2 decades (Hietz et al. 2002; Zotz et al. 2005).

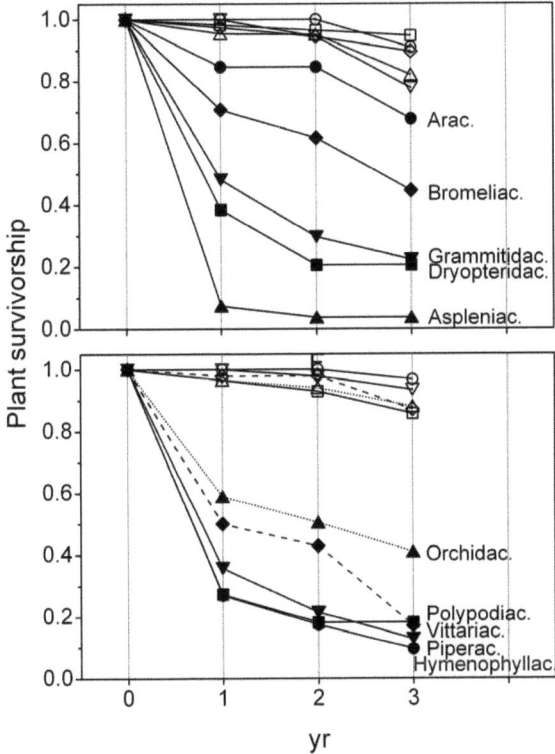

Figure 4. Epiphyte losses unrelated to substrate failure over the course of three years following forest clearance. Filled symbols: IRTs, open symbols: forest; solid lines: ferns, dashed lines: dicots, stippled lines: monocots. Taxa are divided alphabetically between panels.

Thirty percent of the plants lost from intact IRT structures disappeared without traces of a shoot and may have been dislodged from their substrate, related to stronger wind turbulence. Hurricanes can cause high rates of epiphyte dislodgment (Ackerman & Moya 1986; Rodríguez-Robles et al. 1990; Robertson & Platt 2001); however, such extreme wind speeds are not nearly approached at our site (Emck 2007). Not unlikely, many of the plants we classified as disappeared instead fell only after deceasing in situ, or we simply failed to detect the decomposing remains of plants that were never dislodged at all. Indeed, live or freshly dead fallen plants were encountered only in few cases (F. Werner, pers. obs.).

The majority of lost IRT epiphytes (70%) evidently died in situ. Elevated abundance of herbivorous insects has been reported from some forest edges (Edwards-Jones & Brown 1993; Barbosa et al. 2005), but leaf damage from herbivory was generally low in our study (< 5% of leaf area), and did not increase following clear-cutting (F. Werner, unpubl. data). Other biotic edge effects (e.g., competition by matrix species) cannot explain the decline of epiphytes on IRTs either, which strongly suggests a paramount influence of changed abiotic conditions.

Mechanisms—Forest edges experience physical edge effects including increased light levels, wind turbulence, air temperature and lowered air humidity (Murcia 1995; Laurance 2004). Because IRTs essentially constitute miniature forest fragments (Williams-Linera et al. 1995) exposed to multiple edges, it appears safe to assume that physical edge effects affect organisms on IRTs to a particularly severe extent (see Laurance 2007), which was corroborated by microclimate measurements (this study; Werner & Gradstein, submitted b).

In essence, physical edge effects are based on increased wind turbulence and solar radiation, yet these two components may affect epiphytes in a complex and cascading manner. The circumstance that hygrophilous understorey taxa tend to be underrepresented in disturbed habitats has led authors to emphasise the relevance of increased desiccation stress for epiphytes in disturbed habitats (Barthlott et al. 2001; Flores-Palacios & Garcia-Franco 2004; Werner et al. 2005). By transplanting individuals of four species to forest of lower, drier elevations, Nadkarni & Solano (2002) could show clearly that increased temperature and desiccation stress affect even well-established canopy epiphytes. Plants responded with increased leaf mortality and reduced leaf production, yet this field experiment could not distinguish between direct effects of temperature on plant metabolism (e.g., respiration), and indirect effects (via drought stress). Moreover, not only are understorey epiphytes more sensitive to desiccation, but they are also more photolabile and tend to be of more delicate built than canopy taxa (Hietz & Briones 1998, 2001; Griffiths & Maxwell 1999), and therefore also more sensitive to physical edge effects other than desiccation, such as wind turbulence and light. Desiccation stress may thus be only one of several interrelated predictors, and some of the mechanisms by which physical edge effects may affect epiphytes shall be discussed briefly in the following.

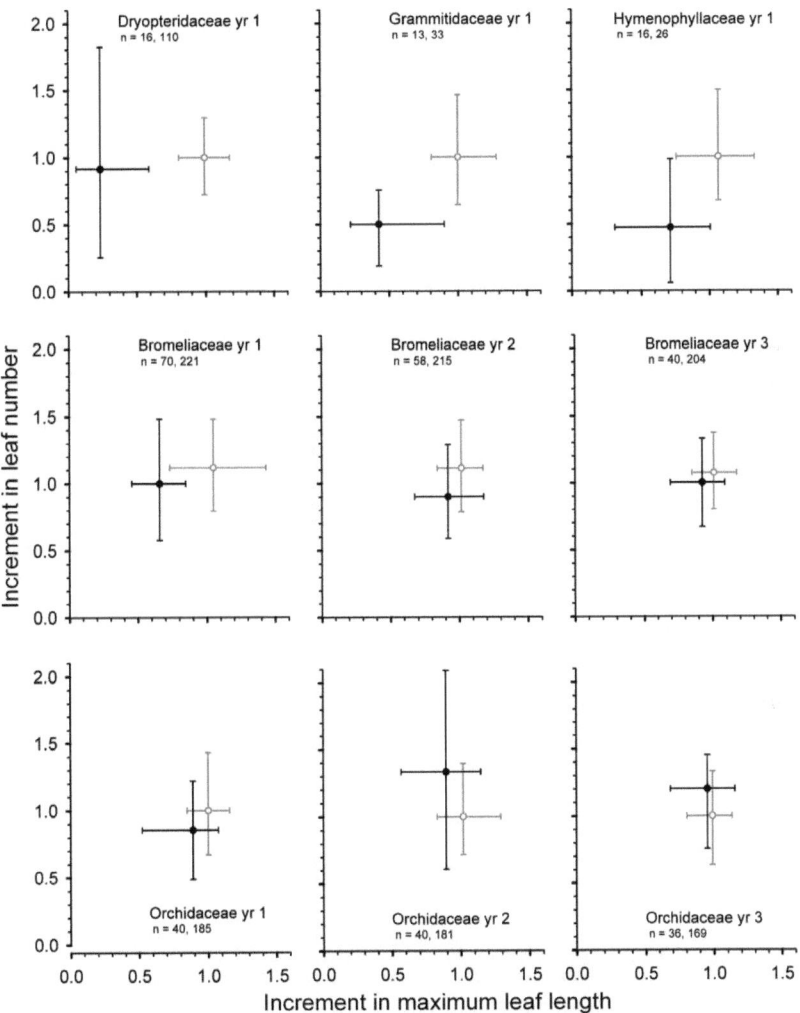

Figure 5. Relative annual increment in leaf number and maximum leaf length. Central dots show median gravity points of IRT plants (black) and control plants in forest (grey), whiskers show the standard deviations of the mean. Given sample sizes are the smallest n of the two parameters in IRT plants and control plants.

Through photoinhibition, intense sunlight can reduce photosynthetic gain, especially under water constraints (Björkman & Powles 1984; Muslin & Homann 1992). Extreme rises in solar radiation can result in light shocks and permanent photodamage. Significant cell damage can also be caused by short-waved radiation (UV) (Kulandaivelu et al. 1997; Björn 2007). Moreover, extended direct sunlight increases the temperature of air, substrate, and plant tissue, which increases respiratory costs and may even cause heat damage to plant tissue in addition to indirect deleterious effects via water relations (Young & Smith 1979; Muslin & Homann 1992; Atkin et al. 2007).

By afflicting injuries to plant structures, increased wind probably contributed to plant decline (compare Sillett et al. 2000), especially in delicate ferns with weak or brittle frond axes such as in *Asplenium* species, which responded particularly drastic to increased exposure. By circulating below-canopy air, even moderate wind turbulence effectively minimises the vertical microclimatic gradient, which guarantees constantly high air humidity and moderate temperatures in the lower strata of intact moist forests (Laurance 2004). Resulting evapotranspiratory losses are further amplified by the (wind-induced) destabilisation and reduction of the leaf boundary layer between tissue surface and atmosphere.

Direct damage through physical edge effects may be magnified by their adverse effects on substrate quality and mycorrhizal symbionts. Intercepting large quantities of moisture, bryophytes are of great importance for the water relations of vascular epiphytes (Dudgeon 1923; Krömer & Gradstein 2003). Three years after clear-felling, we found bryophyte cover on IRTs substantially lower than on forest phorophytes; in fact, most bryophytes on IRTs had already turned tan after only one year (see Fig. 1). This rapid decline following clear-cutting clearly was a consequence of increased physical exposure to atmospheric conditions. Similar to most vascular epiphytes, epiphytic bryophytes are sensitive to desiccation; indeed, their sensitivity may even be far greater related to the lack of a protective cuticle and effective structures for uptake and storage of water (Proctor 2000). Understorey bryophytes further are sensitive to excessive solar radiation (Miyata & Hosokawa 1961). 'Crown humus' is another important resource for epiphytes in montane moist forest which is scanty on IRTs (Werner et al. 2005; F. Werner & C. Gehrig, unpubl. data). Humus accumulation is negatively related to aridity (Birch & Friend 1956), and humus on IRTs is further subjected to erosion by torrential rains and wind. Hence, given the great importance of bryophyte and humus substrates for vascular epiphytes (Dudgeon 1923; Nadkarni 2000; Krömer & Gradstein 2003), direct physical edge effects on vascular epiphytes probably are enhanced by their detrimental effects on substrate quality. The same may apply to mycorrhizal fungi, which form mutualistic relationships with many vascular epiphyte species (Lesica & Antibus 1990; Lehnert 2007), and may be affected by increased physical exposure and substrate decay on IRTs (see Shi et al. 2002). Mycorrhizal fungi provide improve mineral nutrition and water supplies, and improve plant performance under drought stress (Stahl et al. 1998; Augé 2001; Al-Karaki et al. 2004; Subramanian et al. 2006).

To further complicate matters, certain components of edge effects are not necessarily related linearly with epiphyte performance. For instance, the exposure in tree-fall gaps at the same study site did not increase mortality of understorey epiphytes during the first year following experimental felling, and growth was reduced only in few fern taxa (Günter et al., in press). In the following two years, several taxa even showed a clear trend towards enhanced growth in gaps as compared to mature, shady forest stands (F. Werner & P. Ramirez, unpubl. data), suggesting that this moderate level of disturbance rather had positive effects on plant performance.

Temporal and compositional trends—Even if plants were not able to acclimatise to exposure, one would expect mortality on IRTs to decline monotonously with time, given that the most vulnerable individuals are successively excluded. While such a relaxation was evident throughout taxa after the second study year, mortality increased during the following year mortality among several families. Since increased mortality was also observed for several groups of control plants, we attribute it to exceptionally dry conditions in 2006, but future study years will have to confirm this notion.

The high mortality among pteridophytes (87% after three yr) and their pronounced response to exposure in terms of maximal leaf length suggest that they are particularly susceptible to physical edge effects. Several epiphytic fern taxa (Aspleniaceae, Dryopteridaceae, Vittariaceae) also responded with significantly reduced growth when exposed in tree-fall gaps at the same site, whereas other taxa such as bromeliads or aroids did not (Günter et al., in press). These findings are in line with studies that found epiphyte species richness more closely coupled with humidity for pteridophytes than for angiosperms (Johansson 1974; Ibisch 1996; Kessler 2001). Among dicots as a group, mortality was also high (88%) compared to monocots (56%). However, at least in the extreme case treated in our study (i.e., an understorey community subjected to near-maximal exposure), differences in mortality after only three years of observation may eventually turn out to be a sheer matter of response velocity. For instance, the four families suffering the lowest mortality during the first two years (Araceae, Bromeliaceae, Orchidaceae, Piperaceae) did show signs of stabilisation after the 3rd year (Fig. 4). Of the bromeliads censused on IRTs in this study, 98% were species in the genera *Guzmania* and *Pitcairnia*. Despite of their common occurrence in mature forest, representatives of these genera were entirely absent from IRTs 10–30 years after clear-cutting (Werner et al. 2005).

Our study was restricted to lower strata, where the microclimate around IRTs is more strongly altered than in their crowns (F. Werner, unpubl. data). Canopy epiphytes can therefore be expected to respond less drastically to isolation of their hosts in clear-cuts, and a number of xerotolerant upper canopy taxa appear to adjust successfully to growth conditions on IRTs (F. Werner, pers. obs.). However, the overall number of original epiphyte species on IRTs can be presumed to decline negatively exponentially with time. Moreover, recruitment on isolated trees locally is substantially lower than in forest both in terms of abundance and diversity of epiphytes, and strongly biased to xerophilous canopy taxa (Werner & Gradstein, submitted b), suggesting that secondary succession cannot easily offset the losses of original epiphytic flora.

Growth

Our results on plant growth confirm the rapid and general decline of epiphytes following the spatial isolation of their hosts. Throughout taxa, epiphytes responded with decreased leaf length to forest clearance (Fig. 5). Negative growth in maximum leaf length implies both the loss of leaf tissue (partial or complete loss of leaves), and the failure to produce new leaves of similar size (see Fig. 1). As discussed above, leaf tissue is potentially sensitive to a number of environmental factors that change with opening of the canopy. The production of large leaves is a function of both plant vigour and direct environmental constraints. As for the latter, both excessive light levels, temperatures, and effective evapotranspiration discourage the production of large leaves (Gates et al. 1968; Rhizopoulou et al. 1991; Leuschner 2002; Lendzion, in press).

Leaf mortality was generally high in IRT plants (F. Werner, pers. obs.) but often accompanied by a prolific production of smaller, often minute leaves (Fig. 1d). This mode of response explains the combination of great variability and high mean values in leaf number increment exhibited by some plant groups on IRTs (e.g. Dryopteridaceae, Orchidaceae; see Appendix 6, Fig. 5).

Conclusions

Our study provides first evidence from experimental in situ work that increased physical exposure to atmospheric conditions affects abundance and diversity of vascular epiphytes. Throughout the wide range of vascular epiphyte taxa, mortality rates of vascular epiphytes were elevated dramatically following the exposure of their phorophytes in a fresh clear-cut. High mortality of well-established plants can thus explain the marked impoverishment locally found on IRTs after prolonged isolation.

Recruitment limitations may further exacerbate the documented effects of late juvenile and adult mortality following disturbance, particularly at larger time-scales. At our study site, early seedling establishment of epiphytes on isolated trees was 90% lower than in forest (Werner & Gradstein, submitted b). However, growth rates of epiphyte populations have been found to depend almost exclusively on survival, especially of adult plants (Tremblay 1997; Zotz et al. 2005; Zotz & Schmidt 2006; Winkler et al. 2007). Increases in mortality of seedlings on IRTs can further be expected to equal if not exceed those of later life-stages we report, given that epiphyte seedlings are more sensitive to environmental stress (Zotz et al. 2001, Zotz & Hietz 2001). In contrast to other studies (Hietz-Seifert et al.1996; Sillett et al. 2000; Wolf 2005; see also review by Pharo & Zartman 2007), our results therefore suggest that adverse growth conditions often are more limiting to the persistence of epiphyte populations in anthropogenic landscapes than dispersal.

Elevated drought stress is only one out of several complexly interrelated and cascading components of increased physical exposure that may have affected epiphyte performance. Understanding how these factors affect epiphyte performance may be as little trivial as it is simple. For instance, potentially much is to be learned from studies of human disturbance effects on epiphyte responses to global climate change. Just as in disturbed habitats, temperatures and vapour pressure deficit are projected to rise significantly over vast areas with climate change, but levels of light and wind are not. How influential really are climate change-sensitive factors? To which dimension of physical exposure, to which extend and by which means (e.g., pigmentation, anatomy, physiology) can individual plants adjust with time? Descriptive studies from more tropical sites are dearly needed to allow for a generalized picture of disturbance effects beyond local scenarios. However, only their combination with ecological and ecophysiological experiments will lead to a thorough understanding of the processes governing epiphyte communities in non-natural habitats, and eventually permit an extrapolation of accomplished efforts.

REFERENCES

Acebey, A., Gradstein, S.R. & Krömer, T. 2003. Species richness and habitat diversification of corticolous bryophytes in submontane rain forest and fallows of Bolivia. Journal of Tropical Ecology 18: 9–18.

Ackerman, J. & Moya, S. 1986. Hurricane aftermath: resiliency of an orchid-pollinator interaction in Puerto Rico. Caribbean Journal of Science 26: 369–374.

Al-Karaki, G., McMichael, B. & Zak, J. 2004. Field response of wheat to arbuscular mycorrhizal fungi and drought stress. Mycorrhiza 14: 263–269.

Alpert, P. & Oechel, W.C. 1985. Carbon balance limits the distribution of *Grimmia laevigata*, a desiccation-tolerant plant. Ecology 66: 660–669.

Alpert, P. 2000. The discovery, scope, and puzzle of desiccation tolerance in plants. Plant Ecology 151: 5–17.

Andrade, J.L. 2003. Dew deposition on epiphytic bromeliad leaves: an important event in a Mexican tropical dry deciduous forest. Journal of Tropical Ecology 19: 479–488.

Atkin, O.K., Scheurwater, I. & Pons T.L. 2007. Respiration as a percentage of daily photosynthesis in whole plants is homeostatic at moderate, but not high, growth temperatures. New Phytologist 174: 367–380.

Atmar, W. & Patterson, B.D. 1993. The measure of order and disorder in the distribution of species in fragmented habitat. Oecologia 96: 373–382.

Atmar, W. & Patterson, B.D. 1995. The nestedness temperature calculator: a visual basic program, including 294 presence-absence matrices. AICS Research, Univ. Park, NM and The Field Museum, Chicago.

Augé, R.M. 2001. Water relations, drought and vesicular-arbuscular mycorrhizal symbiosis. Mycorrhiza 11: 3–42.

Barbosa, V.S., Leal, I.R., Lannuzzi, L. & Almeida-Cortez, J. 2005. Distribution pattern of herbivorous insects in a remnant of Brazilian Atlantic Forest. Neotropical Entomology 34: 701–711.

Barthlott, W., Schmitt-Neuerburg, V., Nieder, J. & Engwald S. 2001. Diversity and abundance of vascular epiphytes: a comparison of secondary vegetation and primary montane rain forest in the Venezuelan Andes. Plant Ecology 152: 145–156.

Bartoli, C.G., Beltrano, J., Fernández, L.V. & Caldíz, D.O. 1993. Control of the epiphytic weeds *Tillandsia recurvata* and *Tillandsia aëranthos* with different herbicides. Forest Ecology and Management 59: 289–294.

Benavides, A.-M., Wolf, J.H.D. & Duivenvoorden J.F. 2006. Recovery and succession of epiphytes in upper Amazonian fallows. Journal of Tropical Ecology 22: 705–717.

Bendix, J. 2005. Geländeklimatologie. Studienreihe Geographie, Bornträger Verlag, Stuttgart.

Benjamini, Y. & Hochberg, Y. 1995. Controlling the false discovery rate: a practical and powerful approach to multiple testing. Journal of the Royal Statistical Society Series B 57: 289–300.

Benzing, D.H. 1978. Germination and early establishment of *Tillandsia circinnata* Schlecht. (Bromeliaceae) on some of its host trees and other supports in southern Florida. Selbyana 5: 95–106.

Benzing, D.H. 1981. The population dynamics of *Tillandsia circinnata* (Bromeliaceae): cypress crown colonies in southern Florida. Selbyana 5: 256–263.

Benzing D.H. 1990. Vascular epiphytes: general biology and related biota. Cambridge University Press, Cambridge

Benzing, D.H. 1998. Vulnerabilities of tropical forests to climate change: the significance of resident epiphytes. Climate Change 39: 519–540.

Benzing, D.H. & Renfrow, A. 1971. The significance of photosynthetic efficiency to habitat preference and phylogeny among tillandsioid bromeliads. Botanical Gazette 132: 19–30.

Benzing, D.H., Seemann, J. & Renfrow, A. 1978. The foliar epidermis in Tillandsioideae (Bromeliaceae) and its role in habitat selection. American Journal of Botany 65: 359–365.

Bernal, R., Valverde, T. & Hernández-Rosas, L. 2005. Habitat preference of the epiphyte *Tillandsia recurvata* (Bromeliaceae) in a semi-desert environment in Central Mexico. Canadian Journal of Botany 83: 1238–1247.

Birch, H.F. & Friend, M.T. 1956: The organic matter and nitrogen status of East African soils. Journal of Soil Science 7: 156–167.

Björkman, O. & Powles, S. 1984. Inhibition of photosynthetic reactions under water stress: interaction with light level. Planta 161: 490–504.

Björn, L.O. 2007. Stratospheric ozone, ultraviolet radiation, and cryptogams. Biological Conservation 135: 326–333.

Borchsenius, F. 1997. Patterns of plant species endemism in Ecuador. Biodiversity and Conservation 6: 379–399.

Brehm, G., Homeier, J., Fiedler, K., Kottke, I., Illig, J., Nöske, N.M., Werner, F.A. & Breckle, S.-W. In press. Mountain rain forests in southern Ecuador as a hotspot of biodiversity – limited knowledge and diverging patterns. Pp. 15–23 in Beck, E., Bendix, J., Kottke, I., Makeschin, M. & Mosandl, R. (eds.), Gradients in a tropical mountain ecosystem of Ecuador. Ecological Studies Vol. 198, Springer, Berlin.

Brown, S. & Lugo, A.E. 1990. Tropical secondary forests. Journal of Tropical Ecology 6: 1-32.

Bruijnzeel, L.A. 2005. Tropical montane cloud forests: a uniqe hydrological case. Pp. 462–483 in Bonell, M. & Bruijnzeel, L.A. (eds.). Forests, water and people in the humid Tropics. Cambridge University Press. Cambridge.

Burckhardt, H. 1963. Meteorologische Voraussetzungen für Nachtfröste. Pp. 13–81 in Schnelle, F. (ed.). Frostschutz im Pflanzenbau, Vol. I. Bayerischer Landwirtschafts-Verlag, München.

Bush, M.B. 2002. Distributional change and conservation on the Andean flank: a paleoecological perspective. Global Ecology and Biogeography 11: 463–473

Bussmann, R.W. 2001. Epiphyte diversity in a tropical Andean forest – Reserva Biológica San Francisco, Zamora-Chinchipe, Ecuador. Ecotropica 7: 43–59.

Caldíz, D.O. & Fernández, L.V. 1995. The role of the epiphyte weeds *Tillandsia recurvata* and *Tillandsia aëranthos* in native rural and urban forest. International Journal of Ecology and Environmental Sciences 21: 177–197.

Cardoso da Silva, J.M., Uhl, C. & Murray, G. 1996. Plant succession, landscape management, and the ecology of frugivorous birds in abandoned Amazonian pastures. Conservation Biology 10: 491–503.

Carrière, S.M., Mathieuandré, P., Letourmy, P. & McKey, D.B. 2002a. Seed rain beneath remnant trees in a slash-and-burn agricultural system in southern Cameroon. Journal of Tropical Ecology 18: 353–374.

Carrière, S.M., Letourmy, P. & McKey, D.B. 2002b. Effects of remnant trees in fallows on diversity and structure of forest regrowth in a slash-and-burn agriculture system in southern Cameroon. Journal of Tropical Ecology 18: 375–396.

Cascante, A.M. 2006. Epiphytic bromeliad communities during premontane forest succession in Costa Rica. Ph.D. dissertation, Universiteit van Amsterdam.

Cascante-Marín, A., Wolf, J.H.D., Ostermeijer, J.G.B., den Nijs, J.C.M., Sanahuja, O. & Durán-Apuy, A. 2006. Epiphytic bromeliad communities in secondary and mature forest in a tropical montane area. Basic and Applied Ecology. 7: 520–532.

Castañeda, G. 2001. Aves asociadas a plantas epífitas de un bosque nublado en la reserva de bosque integral Otonga, noroccidente de Ecuador. Pp. 327–334 in Nieder, J. & Barthlott, W. (eds.). Results of the Bonn-Quito epiphyte project, funded by the Volkswagen foundation (Vol.1 of 2). Bonn.

Castro Hernández, J.C., Wolf, J.H.D., García-Franco, J.G. & González-Espinosa, M. 1999. The influence of humidity nutrients and light on the establishment of the epiphytic bromeliad *Tillandsia guatemalensis* in the highlands of Chiapas Mexico. Revista Biología Tropical 47: 763–773.

Catling, P.M. & Lefkowitch, L.O. 1989. Associations of Vascular Epiphytes in a Guatemalan Cloud Forest. Biotropica 21: 35–40.

Charles-Dominique, P. 1986. Inter-relations between frugivorous vertebrates and pioneer plants: *Cecropia*, birds and bats in French Guyana. Pp. 119–136 in Estrada, A. & Fleming, T.H. (eds.). Frugivores and seed dispersal. W. Junk, Dordrecht.

Cobb, A.R., Nadkarni, N.M., Ramsay, G.R. & Svoboda, A.J. 2001. Recolonization of bigleaf maple branches by epiphytic bryophytes following experimental disturbance. Canadian Journal of Botany 79: 1–8.

Colwell, R.K. 2005. EstimateS: Statistical estimation of species richness and shared species from samples. Version 7.5. User's Guide and application available from http://purl.oclc.org/estimates (accessed January 2006).

Colwell, R.K., Mao, C.X. & Chang, J. 2004. Interpolating, extrapolating, and comparing incidence-based species accumulation curves. Ecology 85: 2717–2727.

Csintalan, Z., Tacács, Z., Proctor, M.C.F., Nágy, Z. & Tuba, Z. 2000. Early morning photosynthesis of the moss *Tortula ruralis* following summer dew fall in a Hungarian temperate dry sandy grassland. Plant Ecology 151: 151–154.

Davis, S.D., Heywood, V.H., Herrera-MacBryde, O., Villa-Lobos, J. & Hamilton, A.C. (eds.). 1997. Centres of Plant Diversity. A Guide and Strategy for their Conservation. Vol. 3. The Americas. IUCN, Washington, D.C.

Denison, W.C. 1973. Life in tall trees. Scientific American 207: 75–80.

Didham, R.K. & Lawton, J.H. 1999. Edge structure determines the magnitude of changes in microclimate and vegetation structure in tropical forest fragments. Biotropica 31: 17–30.

Dietz, J., Leuschner C., Hölscher, D. & Kreilein, H. 2007. Vertical patterns and duration of surface wetness in an old-growth tropical montane forest, Indonesia. Flora 202: 111–117.

Doyle, G. 2000. Strangler figs in a stand of dry rainforest in the Lower Hunter Valley, NSW. Australian Geographer 31: 251–264.

Drehwald, U. 1995. Epiphytische Pflanzengesellschaften in NO-Argentinien. Ph.D. dissertation. Dissertationes Botanicae 250. J. Kramer, Berlin.

Drehwald, U. 2005. Biomonitoring of disturbance in neotropical rainforests using bryophytes as indicators. Journal of the Hattori Botanical Laboratory 97: 117–126.

Drehwald, U. & Preising, E. 1991. Die Moosgesellschaften Niedersachsens. Moosgesellschaften. Naturschutz und Landschaftspflege 20/9, Niedersächsisches Landesverwaltungsamt für Landschaftspflege, Fachbehörde für Naturschutz, Hannover.

Dudgeon, W. 1923: Successions of epiphytes in the *Quercus incana* forests at Landour, Western Himalayas. Journal of the Indian Botanical Society 3: 270–272.

Dufrêne, M. & Legendre, P. 1997. Species assemblages and indicator species: the need for a flexible asymmetrical approach. Ecological Monographs 67: 345–366.

Duncan, R.S. & Chapman, C.A. 1999. Seed dispersal and potential forest succession in abandoned agriculture in Tropical Africa. Ecological Applications 9: 998–1008.

Dunn, R.R. 2000. Bromeliad communities in isolated trees and three successional stages of an Andean cloud forest in Ecuador. Selbyana 21: 137–143.

Edwards-Jones, G. & Brown, V.K. 1993. Successional trends in insect herbivore population densities: A field test of a hypothesis. Oikos 66: 46–471.

Ek, R.C., ter Steege, H. &. Biesmeijer K.C. 1997. Vertical distribution and associations of vascular epiphytes in four different forest types in the Guianas. Pp. 65-89 in Tropenbos (ed.). Botanical diversity in the tropical rainforest of Guyana. Tropenbos, Utrecht.

Emck, P. 2007. A climatology of South Ecuador. Ph.D. dissertation, Universität Erlangen.

Engwald, S. 1999. Diversität und Ökologie der vaskulären Epiphyten eines Berg- und eines Tieflandregenwaldes in Venezuela. Ph.D. dissertation, Universität Bonn. Libri – books on demand, Hamburg.

Esseen, P.A. 2006. Edge influence on the old-growth forest indicator lichen *Alectoria sarmentosa* in natural ecotones. Journal of Vegetation Science 17: 185–194.

Everts, K.L. & Lacy, M.L. 1990. The influence of dew duration, relative humidity, and leaf senescence on conidial formation and infection of onion by *Alternaria porri*. Phytopathology 80: 1203–1207.

Fajardo, L., Gonzáles, V., Nassar, J.M., Lacabana, P., Portillo, C.A., Carrasquel, F. & Rodríguez, P. 2005. Tropical dry forests of Venezuela: Characterization and current conservation status. Biotropica 37: 531–546.

Fjeldså, J. 1995. Geographical patterns of neoendemic and older relict species of Andean forest birds: the significance of ecologically stable areas. Pp. 89-102 in Churchill, S.P., Balslev, H., Forero, E. & Luteyn, J.L. (eds.). Biodiversity and conservation of neotropical montane forests. Memoirs of the New York Botanical Garden, Bronx, New York.

Fjeldså, J. 2002. *Polylepis* forests – vestiges of a vanishing ecosystem in the Andes. Ecotropica 8: 111–123.

Fleming, T.H., Muchhala, N.C. &. Ornelas P. 2005. New World nectar-feeding vertebrates: community patterns and processes. Pp. 163–186 in Sanchez-Cordero, V. & Medellin R.A. (eds.). Contribuciones Mastozoologicas en Homenaje a Bernardo Villa. UNAM, Conabio.

Flesch, T.K. & Wilson, J.D. 1999. Wind and remnant tree sway in forest cutblocks. I. measured winds in experimental cutblocks. Agricultural and Forest Meteorology 93: 229–242.

Flores-Palacios, A. & García-Franco, J.G. 2001. Sampling methods for vascular epiphytes: their effectiveness in recording species richness and frequency. Selbyana 22:181–191.

Flores-Palacios, A. & García-Franco, J.G. 2004. Effects of isolation on the structure and nutrient content of oak epiphyte communities. Plant Ecology 173: 259–269.

Flores-Palacios, A. & García-Franco, J.G. 2006. The relationship between tree size and epiphyte species richness: testing four different hypotheses. Journal of Biogeography 30: 323–330.

Frahm, J.P. 2001. Biologie der Moose. Spektrum Akademischer Verlag, Berlin.

Freiberg, M. & Freiberg, E. 2000. Epiphyte diversity and biomass in the canopy of lowland and montane forests in Ecuador. Journal of Tropical Ecology 16: 673–688.

Gardner, T.A., Barlow, J., Parry, L.W. & Peres, C.A. 2007. Predicting the uncertain future of tropical forest species in a data vacuum. Biotropica 39:25–30.

Gascon, C., Lovejoy, T.E., Bierregaard, R.O., Malcolm, J.R., Stouffer, P.C., Vasconcelos, H.L., Laurance, W.F., Zimmerman, B., Tocher, M. & Borges, S. 1999. Matrix habitat and species richness in tropical forest remnants. Biological Conservation 91: 223-229.

Gates, D.M. 1968. Transpiration and leaf temperature. Annual Review of. Plant Physiology 19: 211–238.

Gehrig, C. 2005. Human impact on epiphytic biomass in southern Ecuador. Diploma thesis, University of Göttingen.

Gentry, A.H. 1988. Changes in plant diversity and floristic composition on environmental and biogeographical gradients. Annals of the Missouri Botanical Garden 75: 1–34.

Gentry, A.H. 1992. Tropical forest biodiversity: distributional patterns and conservational significance. Oikos 63: 19–28.

Gentry, A.H. 1995. Patterns of floristic composition in neotropical montane forests. Pp. 103–126 in Churchill, S.P., Balslev, H., Forero, E. & Luteyn, J.L. (eds.). Biodiversity and conservation of neotropical montane forests. The New York Botanical Garden, New York.

Gentry, A.H. & Dodson, C.H. 1987a. Diversity and biogeography of neotropical vascular epiphytes. Annals of the Missouri Botanical Garden 74: 205–233.

Gentry, A.H. & Dodson, C.H. 1987b. Contribution of nontrees to species richness of a tropical rain forest. Biotropica 19: 149–156.

Githiru, M., Bennur, L.A. & Ogul C.P.K.O. 2002. Effects of site and fruit size on the composition of avian frugivore assemblages in a fragmented Afrotropical forest. Oikos 96: 320-330.

Gonçalvez, C.N. & Waechter, J.L. 2003. Aspectos florísticos e ecológicos de epífitos vasculares sobre figueiras isoladas no norte da planície costeira do Rio Grande Do Sul. Acta Botanica Brasiliensis 17: 89–100.

Gotelli, N.J. & Entsminger, G.L. 2006. EcoSim: Null models software for ecology. Version 7. Acquired Intelligence Inc and Kesey-Bear, Jericho, VT 05465. Available via http://garyentsmingercom/ecosimhtm (accessed July 2006).

Gradstein, S.R. 1992. The vanishing tropical rain forest as an environment for bryophytes and lichens. Pp. 234–258 in Bates, J.W. & Farmer, A.R. (eds.). Bryophytes and lichens in a Changing Environment. Clarendon Press, Oxford.

Gradstein, S.R. 2006. The lowland cloud forest of French Guiana – a liverwort hotspot. Cryptogamie, Bryologie 27: 141–152.

Gradstein, S.R. & Pócs, T. 1989. Bryophytes. Pp. 311–325 in Lieth, H., & Werger, M.J.A. (eds.). Tropical rainforest ecosystems. Elsevier, Amsterdam.

Gradstein, S.R., Churchill, S.P. & Salazar Allen N. 2001. Guide to the Bryophytes of Tropical America. Memoirs of the New York Botanical Garden 86.

Gradstein, S.R., Nadkarni, N.M., Krömer, T., Holz I. & Nöske, N. 2003. A protocol for rapid and representative sampling of epiphyte diversity of tropical rain forests. Selbyana 24: 87–93.

Graham, E.A. & Andrade, J.L. 2004. Drought tolerance associated with vertical stratification of two co-occurring epiphytic bromeliads in a tropical dry forest. American Journal of Botany 91: 699–706.

Greeney, H.F. 2001. The insects of plant-held waters: a review and bibliography. Journal of Tropical Ecology 17: 241–260.

Griffiths, H. & Maxwell, K. 1999. In memory of C.S. Pittendrigh: Does exposure in forest canopies relate to photoprotective strategies in epiphytic bromeliads? Functional Ecology 13: 15–23.

Guerrón, M., Orellana, A., Loor, A. & Zambrano, J. 2005. Estudio del Bosque Seco en el Bosque Protector Jerusalem. Lyonia 8: 5–18.

Guevara, S. 1995. Connectivity: key in maintaining tropical rain forest landscape diversity: a case study in Los Tuxlas, Mexico. Pp. 63-73 in Halladay, P. & Gilmour, D.A. (eds.). Conserving biodiversity outside protected areas: the role of traditional agro-ecosystems. The IUCN Forest Conservation Programme, New York.

Guevara, S., Purata, S.E. & van der Maarel, E. 1986. The role of remnant forest trees in tropical secondary succession. Vegetatio 66: 77–84.

Guevara, S., Meave, J., Moreno-Casasola, P. & Laborde, J. 1992. Floristic composition and structure of vegetation under isolated trees in Neotropical pastures. Journal of Vegetation Science 3: 655–664.

Guevara, S., Laborde, J. & Sánchez, G. 1998. Are isolated remnant trees in pasture a fragmented canopy? Selbyana 19: 34–43.

Günter, S., Cabrera, O., Weber, M., Stimm, B., Zimmermann, M., Fiedler, K., Knuth, J., Boy, J., Wilcke, W., Iost, S., Makeschin, M., Werner, F.A., Gradstein, S.R. & Mosandl, R. In press. Natural forest management in Neotropical mountain rain forests - an ecological experiment. In Beck, E., Bendix, J., Kottke, I., Makeschin, M. & Mosandl, R. (eds.): Gradients in a tropical mountain ecosystem of Ecuador. Ecological Studies Vol. 198, Springer, Berlin.

Harper, A., MacDonald, S.E., Burton, P.J., Chen, J., Brosowske, K.D., Saunders, S.A., Euskirchen, E., Roberts, D., Jaiteh, M.S. & Esseen, P.-A. 2005. Edge influence on forest structure and composition in fragmented landscapes. Conservation Biology 19: 768–782.

Hartig, K. & Beck, E. 2002. Fire management ruins tropical pastureland: a phytosociological investigation of the agricultural areas in the San Francisco valley, Ecuador. Tagungsband. Abstracts of the 15. annual meeting of the Gesellschaft für Tropenökologie (gtoe), Göttingen.

Hickey, J.E. 1994. A floristic comparison of vascular plant species in Tasmanian old growth mixed forest with regeneration resulting from logging and wildfire. Australian Journal of Botany 42: 383–404.

Hietz, P. 1997. Population dynamics of epiphytes in a Mexican humid montane forest. Journal of Ecology 85: 767–777.

Hietz, P. 1998. Diversity and conservation of epiphytes in a changing environment. Pure and applied Chemistry 70: 2114. Full text published by the International Union of Pure and Applied Chemistry, North Carolina. Available at http://iupac.chemsoc.org/symposia/proceedings/phuket97/hietz.html (accessed May 2006).

Hietz, P. 2005. Conservation of vascular epiphyte diversity in Mexican coffee plantations. Conservation Biology 19: 391–399.

Hietz, P. & Hietz-Seifert, U. 1995a. Composition and ecology of vascular epiphyte communities along an altitudinal gradient in central Veracruz, Mexico. Journal of Vegetation Science 6: 487–498.

Hietz, P. & Hietz-Seifert, U. 1995b. Structure and ecology of epiphyte communities of a cloud forest in central Veracruz, Mexico. Journal of Vegetation Science 6: 719–728.

Hietz, P. & Briones, O. 1998. Correlation between water relations and within-canopy distribution of epiphytic ferns in a Mexican cloud forest. Oecologia 114: 305–316.

Hietz, P. & Briones, O. 2001. Photosynthesis chlorophyll fluorescence and within-canopy distribution of epiphytic ferns in a Mexican cloud forest. Plant Biology 3: 279–287.

Hietz, P., Ausserer, J. & Schindler, G. 2002. Growth, maturation and survival of epiphytic bromeliads in a Mexican humid montane forest. Journal of Tropical Ecology 18: 177–191.

Hietz, P., Buchberger, G. & Winkler, M. 2006. Effect of forest disturbance on abundance and distribution of epiphytic bromeliads and orchids. Ecotropica 12: 103–112.

Hietz-Seifert, U., Hietz, P. & Guevara, S. 1996. Epiphyte vegetation and diversity on remnant trees after forest clearance in southern Veracruz, Mexico. Biological Conservation 75: 103–111.

Hilmo, O. & Såstad, S.M. 2001. Colonization of old-forest lichens in a young and an old boreal *Picea abies* forest: an experimental approach. Biological Conservation 102: 251–259.

Hofstede, R.G.M., Wolf, J.H.D. & Benzing, D.H. 1993. Epiphytic biomass and nutrient status of a colombian upper montane rain forest. Selbyana 14: 37–45.

Höft, R. & Höft, M. 1993: Characteristic epiphyte species of the montane forests in Morobe Province, Papua New Guinea. Proceedings of the Biological Society of New Guinea 1993: 35–39.

Hohnwald; S. 1999. Beiträge zum Mikroklima interandiner Trockentäler Boliviens. Geoökodynamik 20: 221–229.

Holl, K. 1999. Factors limiting tropical rain forest regeneration in abandoned pasture: seed rain, seed germination, microclimate and soil. Biotropica 31: 229–242.

Holz, I. & Gradstein, S.R. 2005a. Cryptogamic epiphytes in primary and recovering upper montane oak forests of Costa Rica – species richness, community composition and ecology. Plant Ecology 178: 89–109.

Holz, I. & Gradstein, S.R. 2005b. Phytogeography of the bryophyte floras of oak forests and páramo of the Cordillera de Talamanca, Costa Rica. Journal of Biogeography 32: 1591–1609.

Homeier, J. 2004. Baumdiversität Waldstruktur und Wachstumsdynamik zweier tropischer Bergregenwälder in Ecuador und Costa Rica. Ph.D. dissertation, Universität Bielefeld. Dissertationes Botanicae 391, Cramer, Berlin.

Homeier, J. & Werner, F.A. In press. Spermatophyta. Pp. 15–58 in Liede-Schumann, S., Breckle, S.-W. (eds.). Checklist Reserva Biológica San Francisco (Prov. Zamora-Chinchipe, S. Ecuador). Ecotropical Monographs 4.

Homeier, J., Dalitz, H., & Breckle, S.-W. 2002. Waldstruktur und Baumartendiversität im montanen Regenwald der Estación Científica San Francisco in Südecuador. Berichte der Reinhold-Tüxen-Gesellschaft 14: 109–118.

Homeier, J., Werner, F.A., Breckle, S.-W., Gradstein, S.R. & Richter, M. In press. Potential vegetation and floristic composition of Andean forests in South Ecuador, with a focus on the Reserva San Francisco. In Beck, E., Bendix, J., Kottke, I., Makeschin, M. & Mosandl, R (eds.). Gradients in a tropical mountain ecosystem of Ecuador. Ecological Studies Vol. 198, Springer, Berlin.

Hubbell, S. 2001. The unified neutral theory of biodiversity and biogeography. Princeton University Press.

Hylander, K. 2005. Aspect modifies the magnitude of edge effects on bryophyte growth in boreal forest. Journal of Applied Ecology 42: 518–525.

Hylander, K. & Hedderson, T.A.J. 2007. Does the width of isolated ravine forests influence moss and liverwort diversity and composition? – A study of temperate forests in South Africa. Biodiversity and Conservation 16: 1441–1458.

Ibisch, P. 1996. Neotropische Epiphytendiversität – das Beispiel Bolivien. Ph.D. dissertation, Universität Bonn. Martina Galunder-Verlag, Wiehl.

INAMHI, 1964–1973. Manual meteorológico. Instituto Nacional de Meteorología y Hidrología del Ecuador, Quito.

Ingram, S. W. & Nadkarni, N. M. 1993. Composition and distribution of epiphytic organic matter in a neotropical cloud forest, Costa Rica. Biotropica 25: 370–383.

Ingram, S.W., Ferrell-Ingram, K. & Nadkarni, N.M. 1996. Floristic composition of vascular epiphytes in a neotropical cloud forest, Monte Verde, Costa Rica. Selbyana 17: 88–103.

Janzen, D.H. 1988. Management of habitat fragments in a tropical dry forest: growth. Annals of the Missouri Botanical Garden 75: 105–116.

Jenik, J. 1973. Root systems of tropical trees. 8. Stilt roots and allied adaptions. Preslia 45: 250–264.

Johansson, D. 1974. Ecology of vascular epiphytes in West African rain forest. Acta Phytogeographica Suecia 59: 1–136. Uppsala.

Kapos, V., Wandelli, E., Camargo, J.L., & Ganade, G. 1997. Edge-related changes in environment and plant responses due to fragmentation in central Amazonia. Pp. 33–44 in Laurance, W.F. & Bierregaard, R.O. Jr. (eds.). Tropical forest remnants: ecology, management, and conservation of fragmented communities. University of Chicago Press, Chicago.

Kelly, D.L. 1985. Epiphytes and climbers of a Jamaican rain forest: vertical distribution, life forms and life histories. Journal of Biogeography 12: 223–241.

Kessler, M. 2001. Pteridophyte species richness in Andean forests in Bolivia. Biodiversity and Conservation 10: 1473–1495.

Kessler, M., Bach, K., Helme, N., Beck, S. & Gonzales, J. 2000. Floristic diversity of Andean dry forests in Bolivia – an overview. Pp. 219–234 in Breckle, S.-W., Schweizer, B. & Arndt, U. (eds.). Results of worldwide ecological studies. Proceedings of the first symposium of the A.F.W. Schimper foundation, Hohenheim, October 1998. Verlag Günter Heimbach, Stuttgart.

King, G.C. & Chapman, W.S. 1983. Floristic composition and structure of a rainforest area 25 yr after logging. Australian Journal of Ecology 8: 415–423.

Kreft, H., Köster, N., Küper, W., Nieder, J. & Barthlott, W. 2004. Diversity and biogeography of vascular epiphytes in Western Amazonia, Yasuní, Ecuador. Journal of Biogeography 31: 1463–1476.

Kress, W.L. 1986. A symposium: the biology of tropical epiphytes. Selbyana 9: 1–22.

Krömer, T., Kessler, M., Gradstein, S.R. & Acebey, A. 2005. Diversity patterns of vascular epiphytes along an elevational gradient in the Andes. Journal of Biogeography 32: 1799–1810.

Krömer, T. & Gradstein, S.R. 2003. Species richness of vascular epiphytes in two primary forests and fallows in the Bolivian Andes. Selbyana 24: 190–195.

Krömer, T. 2003. Diversität und Ökologie der vaskulären Epiphyten in primären und sekundären Bergwäldern Boliviens. Ph.D. dissertation, Universität Göttingen. Cuvillier Verlag, Göttingen.

Kulandaivelu, G., Lingakumar, K. & Premkumar, A. 1997. UV-B radiation. Pp. 41–60 in Prasad, M.N.V. (ed.). Plant ecophysiology. John Wiley & Sons, London.

Kürschner, H. & Parolly, G. 2004. Phytomass and water-storing capacity of epiphytic rain forest bryophyte communities in S Ecuador. Botanische Jahrbücher für Systematik 125: 489–504.

Kürschner, H. 2004. Life strategies and adaptations in bryophytes from the Near and Middle East. Turkish Journal of Botany 28: 73–84.

Laman, T.G. 1995. *Ficus stupenda* germination and seedling establishment in a Bornean rainforest. Ecology 76: 2617–2626.

Larrea, M. 1997. Respuesta de las epífitas vasculares a differentes formas de manejo del bosque nublado, Bosque Protegido Sierrazul, zona de amortiguamento de la Reserva Ecológica Cayambe-Coca, Napo, Ecuador. Pp. 321–346 in Mena, P.A.,. Soldi, A., Alarcón, R., Chiriboga, C. & Suárez, L (eds.). Estudios Biológicos para la conservación. EcoCiencia, Quito.

Larrea, M. 1995. Respuesta de las epífitas vasculares a differentes formas de manejo del bosque nublado, Bosque Protegido Sierrazul, Napo, Ecuador. Unpublished report, EcoCiencia, Quito.

Larson, R.J. 1992. Population dynamics of *Encyclia tampensis* in Florida. Selbyana 13: 50–56.

Laskurain, N.A., Escudero, A., Olano, J.M. & Loidi, J. 2004. Seedling dynamics of shrubs in a fully closed temperate forest: greater than expected. Ecography 27: 650–658.

Laube, S. 2006. Long-term changes of vascular epiphyte dynamics in the tropical lowlands of Panama. Ph.D. dissertation, Universität Kaiserslautern.

Laube, S., & Zotz, G. 2006. Long-term changes of the epiphyte assemblages on the palm *Socratea exorrhiza* in a lowland forest in Panama. Journal of Vegetation Science 17: 307–314.

Laurance, W.F. 2004. Forest-climate interactions in fragmented tropical landscapes. Philosophical Transactions of the Royal Society of London B 359: 345–352.

Laurance, W.F. 2006. Have we overstated the tropical biodiversity crisis? Trends in Ecology and Evolution 22: 65–70.

Laurance, W.F., Ferreira, L.V., Rankin-de Merona, J.M. & Laurance, S. 1998. Rain forest fragmentation and the dynamics of Amazonian tree communities. Ecology 79: 2032–2040.

Laurance, W.F., Williams, G.B., Delamônica, P., Oliveira, A., Lovejoy, T.E., Gascon, C. & L. Pohl. 2001. Effects of strong drought on Amazonian forest fragments and edges. Journal of Tropial Ecology 17: 771–785.

Laurance, W.F. & Peres, C.A. (eds.) 2006. Emerging threats to tropical forests. University of Chicago Press, Chicago.

Laurance, W.F. 2007. Ecosystem decay of Amazonian forest fragments: implications for conservation. Pp. 9–35 in: Tscharntke, T., Leuschner, C., Zeller, M., Guhardja, E. & Bidin, A. (eds.) Stability of tropical rainforest margins. Springer, Berlin.

Lawton, R.O. & Putz, F.E. 1988. Natural disturbance and gap-phase regeneration in a wind-exposed tropical lower montane rain forest. Ecology 69: 764–777.

Lehnert, M. 2007. Diversity and evolution of pteridophytes. Ph.D. dissertation, Universität Göttingen.

Lehnert, M., Kessler, M., Salazar, L.I., Navarrete, H., Werner, F.A & Gradstein, S.R. In press. Pteridophytes. Pp. 59–68 in Liede-Schumann, S., Breckle, S.-W. (eds.). Checklist Reserva Biológica San Francisco (Prov. Zamora-Chinchipe, S. Ecuador). Ecotropical Monographs 4.

Lemmon, P.E. 1957. A new instrument for measuring forest overstory density. Journal of Forestry 55: 667–668.

Lendzion, J. In press. Effects of air humidity on development, physiology and distribution of temperate woodland herbs and tree saplings. Ph.D. dissertation, Universität Göttingen.

Lesica, P. & Antibus, R.K. 1990. The occurrence of mycorrhizae in vascular epiphytes of two Costa Rican rain forests. Biotropica 22: 250–258.

Leuschner, C. 2002. Air humidity as an ecological factor for woodland herbs: leaf water status, nutrient uptake, leaf anatomy, and productivity of eight species grown at low or high vpd levels. Flora 197: 262–274.

Lieberman, D. 1996. Demography of tropical tree seedlings. Pp. 131–138 in Swaine, M.D. (ed.). The ecology of tropical forest tree seedlings. UNESCO Man and Biosphere Series Vol. 18, Paris.

López, R.P. 2003. Phytogeographical relationships of the Andean dry valleys of Bolivia. Journal of Biogeography 30: 1659–1668.

Lovejoy, T,E., Bierregaard, R.O. & Rylands, A.B. 1986. Edge and other effects of isolation on Amazon forest fragments. Pp. 257–285 in Soulé, M.E. (ed.). Conservation Biology: science of diversity. Sinauer, Sunderland, MA.

Lovelock, C.E., Jebb, M. & Osmond, C.B. 1994. Photoinhibition and recovery in tropical plant species – response to disturbance. Oecologia 97: 297–307.

Lowman, M. & Linnerooth W. 1995. Population dynamics of some native Florida epiphytes. II. Mortality after a storm. Journal of the Bromeliad Society 45: 15–17.

Lowman, M., Downey, L., Farres, A. & Mermin, E. 1999. Abundance and mortality of two epiphytic Tillandsias (Bromeliaceae) in a Florida Hammock. Journal of the Bromeliad Society 49: 25–28.

Lugo, A.E. & Scatena, F.N. 1992. Epiphytes and climate change research in the Caribbean: a proposal. Selbyana 13: 121–133.

Madison, M. 1977. Vascular epiphytes: their systematic occurrence and salient features. Selbyana 2: 1–13.

Madison, M. 1979. Distribution of epiphytes in a rubber plantation in Sarawak. Selbyana 5: 107–115.

Malhi, Y. & Phillips, O.L. 2004. Tropical forests and global atmospheric change: a synthesis. Philosophical Transactions of the Royal Society of London B 359: 549–555.

Manning, A.D., Fischer, J. & Lindenmayer, D.B. 2006. Scattered trees are keystone structures – Implications for conservation. Biological Conservation 132: 311–321

Martin, C.E. 1994. Physiological ecology of the Bromeliaceae. Botanical Review 60: 1–82.

Martinez-Yrizar, A., Burquez, A. & Maass, M. 2000. Structure and functioning of tropical deciduous forest in western Mexico. Pp. 19–35 in Robichaux, R.H. & Yetman, D.A. (eds.). The tropical deciduous forest of Alamos, The University of Arizona Press Tucson, Arizona.

Matelson, T.J., Nadkarni, N.M. & Longino, J. Longevity of fallen epiphytes in a neotrpical montane forest. Ecology 74: 265–269.

Mather, P.M. 1976. Computational methods of multivariate analysis in physical geography. J. Wiley and Sons, London.

Mayle, F.E., Beerling, D.J., Gosling, W.D. & Bush, M.B. 2004. Responses of Amazonian ecosystems to climatic and atmospheric carbon dioxide changes since the last glacial maximum. Philosophical Transactions of the Royal Society London B 359: 499–514.

McCune, B. 1993. Gradients in epiphyte biomass in three *Pseudotsuga-Tsuga* forests of different ages in western Oregon and Washington. Bryologist 96: 405–411.

McCune, B. & Mefford, M.J. 1999. PC-ORD. Multivariate analysis of ecological data, version 4. MJM Software Design, Glenden Beach.

McDonnell, M.J. & Stiles, E.W. 1983. The structural complexity of old field vegetation and the recruitment of bird-dispersed plant species. Oecologia 56: 109–116.

McGeoch, M.A. & Chown, S.L. 1998. Scaling up the value of bioindicators. Trends in Ecology and Evolution 13: 47–48.

Mielke, P.W.K., Berry, J. & Medina, J.G. 1982. Climax I and II: distortion resistant residual analyses. Journal of Applied Meteorology 21: 788–792.

Mielke, P.W. Jr. 1984. Meteorological applications of permutation techniques based on distance functions. In: Krishnaiah, P.R. & Sen, P.K. (eds.) Handbook of Statistics, Vol. 4, pp. 818-830. Elsevier, Amsterdam.

Miyata, E. & T. Hosokawa. 1961. Seasonal variation of the photosynthetic efficiency and chlorophyll content of epiphytic mosses. Ecology 42: 766–774.

Moen, J. & Jonsson, B.G. 2003. Edge effects on liverworts and lichens in forest patches in a mosaic of boreal forest and wetland. Conservation Biology 17: 380–388.

Mondragón, D., Durán, R., Ramírez, I. & Olmsted, I. 1999. Population dynamics of *Tillandsia brachycaulos* (Bromeliaceae) in Dzibilchaltun National Park, Yucatán Selbyana 20: 250–255.

Mondragón, D., Calvo-Irabien, L.M & Benzing, D.H. 2004. The basis for obligate epiphytism in *Tillandsia brachycaulos* (Bromeliaceae) in a Mexican tropical dry forest. Journal of Tropical Ecology 20: 97–104.

Murcia, C. 1995: Edge effects in fragmented forests: implications for conservation. Trends in Ecology and Evolution 10: 58–62.

Murphy, P.G. & Lugo, A.E. 1986. Ecology of tropical dry forest. Annual Review of Ecology and Systematics 17: 67–88.

Muslin, E.H. & Homann, P.H. 1992. Light as a hazard for the desiccation-resistant 'resurrection' fern *Pleopeltis polypodioides* L. Plant, Cell and Environment 15: 81–89.

Nadkarni, N.M. 1984. Epiphyte biomass and nutrient capital of a neotropical rainforest. Biotropica 16: 249–256.

Nadkarni, N.M. 1986. The nutritional effects of epiphytes on host trees with special reference to alteration of precipitation chemistry. Selbyana 9: 44–51.

Nadkarni, N.M. 1992. The conservation of epiphytes and their habitats: summary of a discussion at the international symposium on the biology and conservation of epiphytes. Selbyana 13: 140–142.

Nadkarni, N.M. 2000. Colonization of stripped bark surfaces by epiphytes in a lower montane cloud forest, Monteverde, Costa Rica. Biotropica 32: 358–363.

Nadkarni, N.M. & Matelson, T.J. 1989. Bird use of epiphytic resources in neotropical trees. Condor 91: 891–907.

Nadkarni, N.M. & Solano, R. 2002. Potential effects of climate change on canopy communities in a tropical cloud forest: an experimental approach. Oecologia 131: 580–586.

Navarrete, H. 2000. Helechos endémicos del Ecuador: distribución y estado de conservación. Nuestra Ciencia 2: 14–17.

Navratil, S., Brace, L.G., Sauder, E.A. & Lux, S. 1994. Silvicultural and harvesting options to favor immature white spruce and aspen regeneration in boreal mixedwoods. Natural Resources of Canada. Information Report NOR-X-337, Canadian Forest Service Northwest Region, Edmonton, Alberta.

Nepstad, D., Uhl, C. & Serrao, E.A. 1990. Surmounting barriers to forest regeneration in abandoned, highly degraded pastures: a case study from Paragomias, Pará, Brasil. Pp. 215–229 in Anderson, A.B. (ed.). Alternatives to deforestation: steps towards sustainable use of the Amazon rainforest. Columbia University Press, New York.

Nieder, J., Engwald, S., Klawun, M. & Barthlott, W. 2000. Spatial distribution of vascular epiphytes (inclusive hemiepiphytes) in a lowland Amazonian rain forest (Surumoni Crane Plot) in southern Venezuela. Biotropica 32: 385–396.

Nkongmeneck, B.-A., Lowman, M.D. & Atwood, J.T. 2002. Epiphyte diversity in primary and fragmented forests of Cameroon, Central Africa: a preliminary survey. Selbyana 23: 121–130.

Nobel, P.S. 1999. Physicochemical and environmental plant physiology, 2nd ed. Academic Press, San Diego.

Nöske N.M., Hilt, N., Werner, F.A., Brehm, G., Fiedler, K., Sipman, H.J.M. & Gradstein, S.R. In press. Disturbance effects on epiphytes and moths in a montane forest in Ecuador. Basic and Applied Ecology.

Nöske, N. 2005. Effekte anthropogener Störung auf die Diversität kryptogamischer Epiphyten (Flechten, Moose) in Bergwäldern Süd-Ecuadors. Ph.D. dissertation, Universität Göttingen.

Nowicki, C. 2001. Epífitas vasculares de la Reserva Otonga. Epiphytes and canopy fauna of the Otonga rain forest (Ecuador). Pp. 115–160 in Nieder, J. & Barthlott, W. (eds.). Results of the Bonn – Quito epiphyte project, funded by the Volkswagen Foundation, Vol. 2 of 2. Books on Demand (http://www.bod.de), Norderstedt.

Öckinger, E., Niklasson, M. & Nilsson, S.G. 2005. Is local distribution of the epiphytic lichen *Lobaria pulmonaria* limited by dispersal capacity or habitat quality? Biodiversity and Conservation 14: 759–773.

Paoletti, M.G., Taylor, R.A.J., Stinner, B.R., Stinner, D.H. & Benzing, D.H. 1991. Diversity of soil fauna in the canopy and forest floor of a Venezuelan cloud forest. Journal of Tropical Ecology 7: 373–383.

Parolly, G., Kürschner, H., Gradstein, S.R & Schäfer-Verwimp, A. 2004. Cryptogams of the Reserva Biológica San Francisco (Province Zamora-Chinchipe, Southern Ecuador. III. Bryophytes: additions and new species. Cryptogamie, Bryologie 25: 271–289.

Peck, J.E. & McCune, B. 1997. Remnant trees and canopy lichen communities in western Oregon: a retrospective approach. Ecological Applications 7: 1181–1187.

Pedrotti, F., Venanzoni, R. & Suárez Tapia, E. 1988. Comunidades vegetales del Valle de Capinota (Cochabamba – Bolivia). Ecología de Bolivia 11: 25–45.

Perry, D. 1978. A method of access into the crown of emergent and canopy trees. Biotropica 10: 155–157.

Pharo, E. & Zartman, C.E. 2007. Bryophytes in a changing landscape: The hierarchical effects of habitat fragmentation on ecological and evolutionary processes. Biological Conservation 135: 315–325.

Pharo, E.J., Kirkpatrick, J.B., Gilfedder, L., Mendel, L. & Turner, P.A.M. 2005. Predicting bryophyte diversity in grassland and eucalypt-dominated remnants in subhumid Tasmania. Journal of Biogeography 32: 2015–24.

Phillips, O.L. & Malhi, Y. 2005. The prospects for tropical forests in the twenty-first-century atmosphere. Pp. 215-226. in Malhi, Y. & Phillips, O.L. (eds.) Tropical forests and global atmospheric change, Oxford University Press, Oxford.

Pittendrigh, C.S. 1948. The bromeliad-*Anopheles*-malaria complex in Trinidad. I. The bromeliad flora. Evolution 2: 58–89.

Pócs, T. 1980. The epiphytic biomass and its effect on the water balance of two rain forest types in Uluguru Mountains (Tanzania, East Africa). Acta Botanica Academiae Scientarum Hungaricae 26: 143–167.

Porembski, S. & Barthlott, W. 2000. Granitic and gneissic outcrops (inselbergs) as centers of diversity for desiccation-tolerant vascular plants. Plant Ecology 151: 19–28.

Proctor, M.C.F. 2000. The bryophyte paradox: tolerance of desiccation, evasion of drought. Plant Ecology 152: 41–49.

Proctor, M.C.F. & Tuba, Z. 2002. Poikilohydry and homoihydry: antithesis or spectrum of possibilities? New Phytologist 156: 327–349.

Putz, F.E. & Holbrook, N.M. 1989. Notes on the natural history of hemiepiphytes. Selbyana 9: 61–69.

Putz, F.E., Blate, G.M., Redford, K.H., Fimbel, R. & Robinson, J. 2001. Tropical forest management and conservation of biodiversity: an overview. Conservation Biology 15: 7–20.

Ramirez, P.D. 2006. Respuesta de las comunidades de epifitas a la manipulación forestal en un bosque tropical de montaña Austro ecuatoriano. Unpublished licenciatura thesis, Universidad del Azuay.

Rhizopoulou, S., Meletiou-Christou, M.S. & Diamantoglou, S. 1991. Water relations for sun and shade leaves of four Mediterranean evergreen sclerophylls. Journal of Experimental Botany 42: 627–635.

Richardson, D.H.S. 1981. The biology of mosses. Halsted Press, New York.

Richter, M. 1991. Methoden der Klimaindikation durch pflanzenmorphologische Merkmale in den Kordilleren der Neotropis. Die Erde 122: 267–289.

Richter, M. 2003. Using epiphytes and soil temperatures for eco-climatic interpretations in Southern Ecuador. Erdkunde 57: 161–181.

Richter, M. In press. Geographic reasons for an outstanding vascular plant diversity of the area. In Beck, E., Bendix, J., Kottke, I., Makeschin, M. & Mosandl, R. Gradients in a tropical mountain ecosystem of Ecuador. Ecological Studies Vol. 198, Springer, Berlin.

Robertson, K.M., & Platt, W.J. 2001. Effect of multiple disturbances (fire and hurricane) on epiphyte community dynamics in a subtropical forest, Florida, U.S.A. Biotropica 33: 573–582.

Rodríguez-Robles, J.A., Ackerman, J.D. & Meléndez, E.J. 1990. Host tree distribution and hurricane damage to an orchid population at Toro Negro Forest, Puerto Rico. Caribbean Journal of Science 26: 163–164.

Rollenbeck, R, Fabian, P. & Bendix, J. 2005. Precipitation dynamics and chemical properties in tropical mountain forests of Ecuador. Advances in Geosciences 6: 1–4.

Rollenbeck, R., Bendix, J., Fabian, P., Boy, J., Dalitz, H., Emck, P., Oesker, M. & Wilcke, W. 2007. Comparison of different techniques for the measurement of precipitation in tropical montane rain forest regions. Journal of Atmospheric and Oceanic Technology 24: 156–168.

Rollenbeck, R., Bendix, J., Fabian, P. In press. Spatial and temporal dynamics of atmospheric water- and nutrient inputs in tropical mountain forests of southern Ecuador In Bruijnzeel, L.A., Juvik, J., Scatena, F.N., Hamilton, L.S. & Bubb, P. (eds). Forests in the mist. Science for conservation and management of tropical montane cloud forests. University of Hawaii Press, Honolulu.

Rudolph, D., Rauer, G., Nieder, J. & Barthlott, W. 1998. Distributional patterns of epiphytes in the canopy and phorophyte characteristics in a western Andean rain forest in Ecuador. Selbyana, 19: 27–33.

Sánchez-Azofeifa, G.A., Quesara, M., Rodríguez, J.P., Nassar, J.M., Stoner, K.E., Castillo, A., Garvin, T., Zent, E.L., Calvo-Alvarado, J.C., Kalacska, M.E.R., Fajardo, L., Gamon, J.A. & Cueva-Reyes, O. 2005. Research priorities for Neotropical dry forests. Biotropica 37: 477–485.

Sanford, W.W. 1968. Distribution of epiphytic orchids in semi-deciduous tropical forest in southern Nigeria. Journal of Ecology 56: 697–705.

Saunders, D.A., Hobbs, R.J. & Margules, C.R. 1991. Biological consequences of ecosystem fragmentation: a review. Conservation Biology 5: 18–32.

Schimper, A.F.W. 1888. Die epiphytische Vegetation Amerikas. Botanische Mittheilungen aus den Tropen II. Verlag Gustav Fischer, Jena.

Schmidt, G. & Zotz, G. 2002. Inherently slow growth in two Caribbean epiphyte species - a demographic approach. Journal of Vegetation Science 13: 527–534.

Schneider, H., Smith, A.R., Raymond Cranfill, R., Hildebrand, T.J., Haufler, C.H. & Ranker, T.A. 2004. Unravelling the phylogeny of polygrammoid ferns (Polypodiaceae and Grammitidaceae): exploring aspects of the diversification of epiphytic plants. Molecular Phylogenetics and Evolution 31: 1041–1063.

Schrumpf, M., Guggenberg, G., Valarezo, C. & Zech, W. 2001. Tropical montane rain forest soils - development and nutrient status along an altitudinal gradient in the South Ecuadorian Andes. Die Erde 132: 43–59.

Shi, L., Guttenberger, M., Kottke, I. & Hampp, R. 2002. The effect of drought on mycorrhizas of beech (*Fagus sylvatica* L.): changes in community structure, and the content of carbohydrates and nitrogen storage bodies of the fungi. Mycorrhiza 12: 303–311.

Shuttleworth, W.J. 1977. The exchange of wind-driven fog and mist between vegetation and the atmosphere. Boundary-Layer Meteorology 12: 463–489.

Sillett, S.C., Gradstein, S.R. & Griffin, D. 1995. Bryophyte diversity of *Ficus* tree crowns from cloud forest and pasture in Costa Rica. Bryologist 98: 251–260.

Sillett, S.C., McCune, B., Peck, J.E., Rambo, T.R. & Ruchty, A. 2000. Dispersal limitations of epiphytic lichens result in species dependent on old-growth forests. Ecological Applications 10: 789–799.

Sipman, H.J. & Harris, R.C. 1989. Lichens. Pp. 303–310 in Lieth, H. & Werger M.J.A. (eds.). Tropical rainforest ecosystems. Elsevier, Amsterdam.

Snäll, T., Ribeiro, P.J. & Rydin, H. 2003. Spatial occurrence and colonizations in patch-tracking metapopulations: local conditions versus dispersal. Oikos 103: 566–578.

Snäll, T., Ehrlén, J. & Rydin, H. 2005. Colonization-extinction dynamics of an epiphyte metapopulation in a dynamic landscape. Ecology 86: 106–115.

Stahl, P.D., Schuman, G.E., Frost, S.M. & Williams, S.E. 1998. Arbuscular mycorrhizae and water stress tolerance of Wyoming big sagebrush seedlings. Soil Science Society of America Journal 62: 1309–1313

Stuntz, S., Ziegler, C., Simon, U. & Zotz, G. 2002. Diversity and structure of the arthropod fauna within three canopy epiphyte species in central Panama. Journal of Tropical Ecology 18: 161–176.

Subramanian, K.S., Santhanakrishnan, P. & Balasubramanian, P. 2006. Responses of field grown tomato plants to arbuscular mycorrhizal fungal colonization under varying intensities of drought stress. Scientia Horticulturae 107: 245–253.

Sudgen, A.M. & Robbins, R.J. 1979. Aspects of the ecology of vascular epiphytes in Colombian cloud forests, 1. The distribution of the epiphytic flora. Biotropica 11: 173–188.

Tanner, E.V.J. 1980. Studies on the biomass and productivity in a series of montane rain forests in Jamaica. The Journal of Ecology 68: 573–588.

ter Steege, H. & Cornelissen, J.H.C. 1989. Distribution and ecology of vascular epiphytes in lowland rain forest of Guyana. Biotropica 21: 331–339.

Thornthwaite, C.W. & Mather, J.R. 1957. Instructions and tables for computing potential evapotranspiration and the water balance. Drexel Institute of Technology, Laboratory of Climatology, Publications in Climatology 10: 185–311.

Todzia, C. 1986. Growth habits, host tree species, and density of hemiepiphytes on Barro Colorado Island, Panama. Biotropica 18: 22–27.

Tremblay, R.L. 1997. *Lepanthes caritensis*, an endangered orchid: no sex, no future? Selbyana 18: 160–166.

Tremblay, R.L., Zimmerman, J.K., Lebrón, L., Bayman, P., Sastre, I., Axelrod, F. & Alers-García, J. 1998. Host specificity and low reproductive success in the rare endemic Puerto Rican orchid *Lepanthes caritensis*. Biological Conservation 85: 297–304.

Turner, I.M., Tan, H.T.W., Wee, Y.C., Ibrahim, A.B., Chew, P.T. & Corlett, R.T. 1994. A study of plant species extinction in Singapore: lessons for the conservation of tropical biodiversity. Conservation Biology 8: 705–712.

UNEP 1992. World atlas of desertification. Edward Arnold, London.

Valencia, R., Pitman, N., León-Yánez, S. & Jørgensen, P.M. 2000. Libro rojo de las plantas endemicas del Ecuador. Herbario QCA, Universidad Católica, Quito.

Vance, E.D. & Nadkarni, N.M. 1990. Microbial biomass and activity in canopy organic matter and the forest floor of a tropical cloud forest. Soil Biology and Biochemistry 22: 677–684.

Vandunné, H.J.F. 2002. Effects of the spatial distribution of trees, conspecific epiphytes and geomorphology on the distribution of epiphytic bromeliads in a secondary montane forest (Cordillera Central, Colombia). Journal of Tropical Ecology 18: 193–213.

Veneklaas, E.J., Zagt, R.J., van Leerdam, A., van Ek, R., Broekhoven, A.J. & van Gederen, M. 1990. Hydrological properties of the epiphyte mass of a montane tropical rain forest, Colombia. Vegetatio 89: 183–192.

Vogelmann, H.W. 1973. Fog precipitation in the cloud forests of eastern Mexico. Bioscience, 23: 96–100.

Webster, G.L. & Rhode, R.M. 2001. Plant diversity of an Andean cloud forest: inventory of the vascular plants of Maquipucuna, Ecuador. University of California Publications in Botany 82, Berkeley.

Werner, F.A., Homeier, J. & Gradstein, S.R. 2005. Diversity of vascular epiphytes on isolated remnant trees in the montane forest belt of southern Ecuador. Ecotropica 11: 21–40.

Werner, F.A. & Gradstein, S.R. In press. Spatial distribution and abundance of epiphytes across a gradient of human disturbance in an Interandean dry valley, Ecuador. Journal of Vegetation Science.

Werner, F.A. & Gradstein, S.R. Submitted. Seedling establishment of vascular epiphytes on isolated and enclosed forest trees in an Andean landscape, Ecuador.

Werth, S., Wagner, H.H., Gugerli, F., Holderegger, R., Csencsics, D., Kalwu, J.W. & Scheidegger, C. 2007. Quantifying dispersal and establishment limitations in a population of an epiphytic lichen. Ecology 87: 2037–2046.

Whittaker, R.H. 1967. Gradient analysis of vegetation. Biological Reviews 42: 207–264.

Williams-Linera, G, Sosa V. & Platas, T. 1995. The fate of epiphytic orchids after fragmentation of a Mexican cloud forest. Selbyana 16: 36–40.

Williams-Linera, G. 1990. Vegetation structure and environmental conditions of forest edges in Panama. Journal of Ecology 78: 356–373.

Wilson, T.B., Bland, W.L. & Norman, J.M. 1999. Measurement and simulation of dew accumulation and drying in a potato canopy. Agricultural and Forest Meteorology 93: 111–119.

Winkler, M., Hülber, K. & Hietz, P. 2005. Effect of canopy position on germination and survival of epiphytic bromeliads in a Mexican humid montane forest. Annals of Botany 95: 1039–1047.

Winkler, M., Hülber, K. & Hietz, P. 2007. Population dynamics of epiphytic bromeliads: Life strategies and the role of host branches. Basic and Applied Ecology 8:183–196.

Wittich, B. 2007. Einfluß von experimentell manipulierter Luftfeuchte und Strahlung auf Wachstum und Morphologie von Waldbodenpflanzen. Unpublished diploma thesis, Universität Göttingen.

Wolf, J.H.D. 1993. Diversity patterns and biomass of epiphytic bryophytes and lichens along an altitudinal gradient in the northern Andes. Annals of the Missouri Botanical Garden 80: 928–960.

Wolf, J.H.D. 1995. Non-vascular epiphyte diversity patterns in the canopy of an upper montane rain forest (2550-3670 m), Central Cordillera, Colombia. Selbyana 16: 185–195.

Wolf, J.H.D. 2005. The response of epiphytes to anthropogenic disturbance of pine-oak forests in the highlands of Chiapas, Mexico. Forest Ecology and Management 212: 376–393.

Wright, D.H., Patterson, B.D., Mikkelson, G., Cutler, A.H. & Atmar, W. 1998. A comparative analysis of nested subset patterns of species composition. Oecologia 113: 1–20.

Yeaton, R.I. & Gladstone, D.E. 1982. The pattern of colonization of epiphytes on calabash trees (*Crescentia alata* HBK.) in Guanacaste Province, Costa Rica. Biotropica 14: 137–140.

Young, D.R. & Smith, W.K. 1979. Influence of sunflecks on the temperature and water relations of two subalpine understory congeners. Oecologia 43: 195–205.

Zahawi, R.A. & Augspurger, C.K. 2006. Tropical forest restoration: tree islands as recruitment foci in degraded lands of Honduras. Ecological Applications 16: 464–478.

Zartman, C.E. 2003. Habitat fragmentation impacts on epiphyllous bryophyte communities in the forests of central Amazonia. Ecology 84: 948–54.

Zartman, C.E. & Nascimento, H.E.M. 2006. Are habitat-tracking meta-communities dispersal-limited? Inferences from abundance-occupancy patterns of epiphylls in Amazonian forest fragments. Biological Conservation 127: 46–54.

Zimmerman, J.K. & Olmsted, I.C. 1992. Host Tree Utilization by Vascular Epiphytes in a Seasonally Inundated Forest (Tintal) in Mexico. Biotropica 24: 402–407.

Zimmerman, J.K., Pascarella, J.B. & Aide, T.M. 2000. Barriers to forest regeneration in an abandoned pasture in Puerto Rico. Restoration Ecology 8: 350–360.

Zotz, G. 1995. How fast does an epiphyte grow? Selbyana 16: 150–154.

Zotz, G. 1998. Demography of the epiphytic orchid, *Dimerandra emarginata*. Journal Tropical Ecology 14: 725–741.

Zotz, G. 1999. Altitudinal changes in diversity and abundance of non-vascular epiphytes in the Tropics – an ecophysiological explanation. Selbyana 20: 256–260.

Zotz, G. 2004a. Long-term observation of the population dynamics of vascular epiphytes. Pp. 97–117 in Breckle, S.-W., Schweizer, B. & Fangmeier, A. (eds.). Results of worldwide ecological studies. Proceedings of the 2nd Symposium of the A.F.W. Schimper-Foundation.

Zotz, G. 2004b. Growth and survival of the early growth stages of the heteroblastic bromeliad *Vriesea sanguinolenta*. Ecotropica 10: 51–57.

Zotz, G. & Tyree, M.T. 1996. Water stress in the epiphytic orchid, *Dimerandra emarginata* (G. Meyer) Hoehne. Oecologia 107: 151–159.

Zotz, G., Hietz, P. & Schmidt, G. 2001. Small plants, large plants – the importance of plant size fort he physiological ecology of vascular epiphytes. Journal of Experimental Botany 52: 2051–2056.

Zotz, G. & Hietz, P. 2001. The physiological ecology of vascular epiphytes: current knowledge, open questions. Journal of Experimental Botany 52: 2067–2078.

Zotz, G., Laube, S. & Schmidt, G. 2005. Long-term population dynamics of the epiphytic bromeliad, *Werauhia sanguinolenta*. Ecography 28: 806–/814.

Zotz, G. & Schmidt, G. 2006. Population decline in the epiphytic orchid *Aspasia principissa*. Biological Conservation 129: 82–90.

ABSTRACT

While tropical deforestation proceeds at high pace, conservation biologists remain divided over the extent to which anthropogenic habitats will be able to offset the loss of biodiversity from primary forests. Case studies addressing disturbance effects on epiphyte assemblages have yielded contrasting patterns, and underlying mechanisms remain unclear. At present, one of the questions most widely debated is the importance of growth conditions vs. dispersal limitations for the persistence of epiphyte populations following disturbance. The present dissertation aims at contributing to our understanding of the processes that determine epiphyte diversity in anthropogenic landscapes, focussing primarily on assemblages of vascular epiphytes on isolated trees. Isolated trees offer an excellent model system since they constitute the smallest possible forest fragment, exposed to multiple edge effects and dispersal limitations.

Field work was done from 2003–2006 at two lower montane sites in the Ecuadorean Andes: (1) moist forest (2200 mm rain/yr; 1800–2200 m elevation) and adjacent matrix habitat around Estación Científica San Francisco (ECSF) on the eastern Andean slope (Zamora-Chinchipe Province), and (2) perarid Interandean forest (530 mm rain/yr; 2300m elevation) and adjacent pastures at Bosque Protector Jerusalén (Pichincha Province).

At the moist forest site (ECSF), I studied the diversity of vascular epiphytes on remnant trees ca. 10–30 years after their isolation in pastures (chapter 2). The objective of this study was to document the importance of remnant trees for the survival of vascular epiphyte communities following forest clearance. Trees were divided into five zones following Johansson (1974) and climbed with the single rope technique. Abundance and diversity of epiphytes were significantly lower on remnant trees than on forest trees. Several families rich in xerophilous taxa (Bromeliaceae, Orchidaceae, Piperaceae, Polypodiaceae) were relatively well represented on remnant trees in terms of species richness and abundance, whereas essentially hygrophilous families such as Dryopteridaceae, Ericaceae, Grammitidaceae and Hymenophyllaceae were poorly represented or absent. Impoverishment was greatest on the stem base and in the outer crown, and least in the inner crown of the host trees. These results suggest that the observed impoverishment on isolated trees is closely related to microclimatic changes.

Experimental work done at the same site gave some insight into the mechanisms behind the pronounced impoverishment of epiphyte assemblages. In one field trial (chapter 6) I studied abundance, diversity and floristic composition of epiphyte seedling establishment on isolated and adjacent forest trees. All vascular epiphytes were removed from plots on the trunk bases of a common tree species (*Piptocoma discolor*). Newly-established epiphyte seedlings were recorded after two years, and their survival after another year. Seedling density and the number of taxa (families and genera, respectively) per plot were significantly reduced on isolated trees relative to forest trees, as was the rarefied total number of epiphyte taxa. Seedling assemblages on trunks of forest trees were dominated by hygrophilous understorey ferns whereas assemblages on isolated trees were dominated by xerophilous canopy taxa. Colonisation probability was significantly higher for plots closer to forest but not for plots with greater canopy or bryophyte cover. Seedling mortality on isolated trees was significantly higher for hygrophilous than for xerophilous taxa. These results show that altered recruitment can explain the long-term impoverishment of post-juvenile epiphyte assemblages on isolated remnant trees. Altered establishment is attributed to a combination of dispersal constraints and the changed microclimate that was documented by measurements of temperature and humidity. Although isolated trees in anthropogenic landscapes are key structures for the maintenance of forest biodiversity in many aspects, my results show that their value for the conservation of epiphyte diversity can be limited. Abiotic seedling requirements may increasingly constitute a bottleneck for the persistence of vascular epiphytes in the face of ongoing habitat alteration and atmospheric warming.

In a second field trial at the same site (chapter 7) I studied the response of assemblages of well-established vascular epiphytes to severe forest disturbance. Individual plants were marked on isolated remnant trees (up to 5 m above ground) in a fresh clearing and in undisturbed forest (controls), and their growth and survival was followed during three consecutive years. Wind-throw and branch breakage caused the loss of 24% of plants from isolated trees but only of < 1% in the forest. Epiphyte mortality on the remaining, intact phorophytes was substantially higher on remnant trees than on forest trees, averaging 72% vs. 11% after 3 years. Mortality on isolated trees was greatest after the first year (52%) and among ferns and dicots, and lowest after the second year (20%) and among monocots (aroids, bromeliads, orchids). Mortality differed significantly between taxa but was elevated significantly for most taxa after the first year relative to forest levels. Plants that survived on isolated trees commonly showed a marked decrease in maximum leaf length. The annual increment in leaf number varied more widely, both between and within epiphyte families. This study provides first experimental, field-based evidence that increased physical exposure affects the performance of well-established vascular epiphytes. The results suggest that growth conditions may often be a more influential driver of vascular epiphyte diversity in disturbed habitats than dispersal. It is argued that elevated desiccation stress is only one of several mechanisms by which increased physical exposure may limit epiphyte performance in disturbed habitats.

The two trials showed that the marked impoverishment of vascular epiphyte assemblages at the study site results from both lowered recruitment and survival of established plants, and that mid-term effects of changes in growth conditions can be drastic.

At the dry forest site (Bosque Protector Jerusalén) I studied the development of epiphyte diversity and abundance across a disturbance gradient (chapters 3 and 4). Epiphytic bryophytes and vascular plants were sampled on 100 trees of the dominant tree species, *Acacia macracantha*, in five habitats: closed-canopy mixed and pure acacia forest (old secondary), forest edge, young semi-closed secondary woodland, and isolated trees in grasslands. Host trees were divided into four zones (cf. Johansson 1974). Species density and surface cover of vascular epiphytes were determined for each zone, bryophytes were only sampled in the inner crown. Total species richness in forest edge and on isolated trees was significantly lower than in closed forest types. Species density of vascular epiphytes did not differ significantly between habitat types. Species density of bryophytes, in contrast, was significantly lower in forest edge and on isolated trees than in closed forest. Surprisingly, edge habitat showed greater impoverishment than semi-closed woodland and similar floristic affinity to isolated trees and to closed forest types. Assemblages were significantly nested; habitat types with major disturbance held only subsets of the closed forest assemblages, indicating a gradual reduction in niche availability.

This conclusion is lent further support by patterns of epiphyte abundance. Bromeliads were xerophilous and early-successional, being most abundant in the middle crown, whereas ferns were skiophilous and late-successional, being essentially restricted to the inner crown. While the former were significantly more abundant on isolated trees than in closed forest, the latter were restricted to closed forest and semi-closed woodland. Bryophyte cover was significantly lower on isolated trees and in forest edge than in closed forest. Total species density and covers of epiphytes on isolated trees were not related to the distance to forest, but total species density and bryophyte cover were closely correlated with crown closure as a measure of canopy integrity.

These results suggest that microclimatic changes are key determinants of the observed impoverishment of epiphyte assemblages following disturbance, and that epiphytic cryptogams are sensitive indicators of microclimate and human disturbance in montane dry forests. Patterns of abundance and diversity further suggest that the 'similar gradient hypothesis' (McCune 1993) applies to tropical dry forest assemblages of epiphytes. The substantial impoverishment of edge habitat underlines the need for fragmentation studies on epiphytes elsewhere in the Tropics.

Case studies on vascular epiphyte assemblages have failed to find a common, general response to human disturbance. In chapter 5, a hypothesis is proposed and explored that aims to explain and unite these conflicting patterns in the light of a non-linear effect of local climate: while the diverse assemblages of moist, slightly seasonal forests rely on the integrity of forest canopies during droughts, this dependency may be less pronounced in aseasonally wet or distinctly dry forests. This hypothesis is supported by the results of a literature survey.

RESUMEN

Mientras la deforestación tropical sigue avanzando a gran paso, los biólogos conservacionistas permanecen divididos tratando de definir que habitas antropogénicos serán capaces de contrarrestar la pérdida de biodiversidad de los bosques primarios.

Casos de estudio refiriéndose a los efectos de la alteración de conjuntos de epífitas han dado como resultado patrones contrastantes, y los mecanismos subyacentes son inciertos. En la actualidad, una de las preguntas mas ampliamente debatida es la importancia de las condiciones de crecimiento vs. limitaciones en la dispersión para la persistencia de las poblaciones de epífitas posteriores a perturbación. La presente disertación apunta a contribuir en el entendimiento de los procesos que determinan la diversidad epifítica en paisajes antropogénicos, enfocándose primariamente en conjuntos de epífitas vasculares y árboles aislados, los cuales pueden actuar como un excelente sistema modelo. Arboles aislados constituyen el fragmento de bosque más pequeño posible, expuesto a multiples efectos de borde y colonización restringida.

El trabajo de campo se llevó a cabo entre 2003–2006 en dos sitios montanos bajos en los Andes Ecuatorianos: (1) Bosque humedo (2200 mm lluvia/año; 1800–2200 m de altitud) y un hábitat matriz adyacente alrededor de la Estación Científica San Francisco (ECSF) en la ladera andina este (Provincia Zamora-Chinchipe) y (2) bosque Interandino perárido (530 mm lluvia-año; 2300 m de altitud) y las pasturas adyacentes al Bosque Protector Jerusalén (provincia de Pichincha).

En el sitio de bosque nublado (ECSF), se estudió la diversidad de las epífitas vasculares en árboles remanentes aislados en pasturas desde hace ca. 10–30 años (Capítulo 2). El objetivo de este estudio fue documentar la importancia de los árboles remanentes para la sobre vivencia de las epífitas vasculares posteriores del clareo de bosque. Los árboles fueron divididos en cinco zonas siguiendo a Johansson (1974) y trepados usando la técnica de cuerda simple ('single rope technique'). La abundancia y diversidad de epífitas fueron significativamente mas bajos en árboles remanentes en términos de riqueza de especies y abundancia. Algunas familias ricas en especies xerófilas (Bromeliaceae, Orchidaceae, Piperaceae, Polypodiaceae) se mantuvieron representadas relativamente bien, mientras que familias esencialmente hidrófilas tales como Dryopteridaceae, Ericaceae, Grammitidaceae y Hymenophyllaceae fueron pobremente representadas o ausentes. El empobrecimiento fue mayor en la base del tronco y en el exterior de la copa, y menor en la copa interna de los árboles hospederos. Estos resultados sugieren que el empobrecimiento observado en árboles aislados está cercanamente relacionado con los cambios del microclima.

Trabajo experimental realizado en el mismo sitio proporcionó alguna sobre los mecanismos detrás de este pronunciado empobrecimiento de conjuntos de epífitas. En una prueba de campo (Capítulo 6) se estudió la abundancia, diversidad y composición florística del establecimiento de plántulas de epífitas en árboles aislados y de bosque. Todas las epífitas vasculares fueron removidas de cuadrantes en las bases de los troncos de un arbol comun (*Piptocoma discolor*). Nuevas plántulas de epífitas establecidas fueron registradas después de dos años y su sobrevivencia después de otro año. Densidad de plántulas y el numero de taxones (familias y géneros, respectivamente) por cuadrante se vieron significantemente reducidos en árboles aislados en relación con los árboles de bosque, así como el número total rarificado de taxones de epífitas. Los conjuntos de plántulas en troncos de árboles de bosque fueron dominados por helechos hidrófilos de sotobosque, mientras que los conjuntos en árboles aislados fueron dominados por taxones xerofíticos de dosel. La probabilidad de colonización fue significativamente alta para los cuadrantes cercanos al bosque pero no para los cuadrantes con una mayor cobertura de dosel y de briofitas. La mortalidad en los árboles aislados fue significativamente alta para taxones hidrófilos que para taxones xerófilos. Estos resultados muestran que un reclutamiento alterado puede explicar el empobrecimiento a largo plazo de conjuntos de epífitas post juveniles en árboles aislados remanentes. El establecimiento alterado es atribuido a una combinación de las restricciones en la dispersión y el riguroso microclima documentado mediante medidas de temperatura y humedad. Aunque los árboles aislados en paisajes antropogénicos son estructuras clave para el mantenimiento de la biodiversidad en muchos aspectos, los resultados muestran que su valor para la conservación de epífitas puede ser limitado. Los requerimientos abióticos altos de crecimiento de las plántulas podrían formar un 'cuello de botella' para la persistencia de epífitas vasculares ante de alteración de hábitat y calentamiento atmosférico progresivo.

En el segundo experimento de campo en el mismo sitio (Capítulo 7), se estudió la respuesta de conjuntos de epífitas vasculares bien establecidos a una severa alteración del bosque. Plantas individuales (hasta 5 m de altura) fueron marcadas en árboles aislados remanentes en un claro de bosque reciente y en un bosque no alterado (controles) y monitoreadas durante 3 años consecutivos. Caída por viento y rotura de ramas sumaron un 24% de pérdida de plantas de árboles aislados, pero menos de 1% en el bosque. Para las epífitas vasculares inafectadas por la caída de sus arboles hospederos y rotura de ramas, la mortalidad fue substancialmente alta en árboles aislados, promediadon un 72% en comparación con un 11% en el bosque, después de 3 años. La mortalidad en árboles aislados fue mayor después del primer año (52%) y entre helechos y dicotiledóneas y el más bajo después del segundo año (20%) y entre las monocotiledóneas (aráceas, bromelias, orquídeas). Sin embargo, la mortalidad fue elevada significativamente a través de los taxones, y las familias que sufrieron pérdidas relativamente bajas durante los primeros 2 años tendieron a sufrir altas pérdidas en los siguientes años. Las plantas sobrevivientes mostraron una reducción en el largo máximo de hoja, pero esta tendencia no fue generalmente seguida por una reducción en el número de hojas. Estos resultados proveen primera evidencia experimental de observaciones de campo de que el incremento de exposición física afecta el desempeño de las epífitas vasculares bien establecidas y sugiere que las condiciones de crecimiento a menudo pueden ser un predictor más influyente de la diversidad de epífitas vasculares en hábitat alterados que la dispersión. Se arguye que el estrés a elevada disecación es solo uno de los muchos mecanismos por el cual el incremento a exposición física limitaría el desempeño de las epífitas en habitas alterados.

Estos dos experimentos muestran así que el marcado empobrecimiento de los conjuntos de epífitas vasculares en el sitio de estudio es resultado tanto de un reclutamiento decaído como de la sobrevivencia de las plantas establecidas, y que efectos a medio término de los cambios en las condiciones ambientales de crecimiento pueden ser drásticos.

En el bosque seco (Bosque Protector Jerualén), se estudió la diversidad y abundancia de epífitas a lo largo de un gradiente de perturbación (Capítulos 3 y 4). Briofitas y plantas vasculares fueron muestreadas en 100 árboles de la especie arbórea dominante, *Acacia macracantha*, en cinco habitas: bosque mezclado de dosel cerrado y bosque de acacia (secundario maduro), borde del bosque, bosque secundario joven semi-cerrado y árboles aislados en pasturas. Los árboles hospederos fueron divididos en cuatro zonas (cf. Johansson 1974). Densidad de especies y cobertura de la superficie de las epífitas vasculares fueron determinadas para cada zona, las briofitas fueron solo muestreadas en la copa interior. La riqueza total de especies en el borde de bosque y en árboles aislados fue significativamente baja en comparación con los tipos de bosque cerrado. La densidad de especies de epífitas vasculares no difirieron significativamente entre tipos de habitas. La densidad de especies de briofitas, en cambio, fue significativamente más baja en el borde de bosque y árboles aislados que en el bosque cerrado. Sorprendentemente, el hábitat de borde mostró un mayor empobrecimiento que el bosque semi-cerrado y afinidad florística similar con los árboles aislados y los tipos de bosque cerrado. Los conjuntos fueron a agrupados significativamente, tipos de habitas con una mayor perturbación sostuvieron solo sub-grupos de los conjuntos de bosque cerrado, indicando una reducción gradual de la disponibilidad de nicho.

Esta conclusión es extensamente apoyada por los patrones de abundancia de epífitas. Las bromelias se presentaron como xerofíticas y tempranas sucesoras, siendo más abundantes en la copa media, mientras que los helechos prefirieron la sombra y fueron sucesores tardíos, estando restringidos esencialmente a la copa interna. Mientras que las primeras fueron significativamente más abundantes en árboles aislados que en bosque cerrado, las últimas se restringieron al bosque cerrado y al bosque semi-cerrado. La cobertura de briofitas fue significativamente más baja en árboles aislados y en borde de bosque que en bosque cerrado. La distancia al bosque no tuvo efectos en la densidad total de especies o coberturas de epífitas en árboles aislados, pero la densidad total de especies y la cobertura de briofitas estuvieron cercanamente correlacionadas con el cierre de la copa como medida de la integridad del dosel.

Estos resultados sugieren que los cambios microclimáticos son una clave determinante del empobrecimiento observado de los conjuntos de epífitas que ocurren posteriormente a perturbación y que las criptógamas epífitas son indicadores sensibles de la perturbación humana y microclimática en bosque seco montano. Patrones de abundancia y diversidad sugieren que la hipótesis de gradiente similar ('similar gradient hypothesis') (McCune 1993) se aplica a conjuntos de bosque seco tropical. El empobrecimiento substancial del hábitat de borde subraya la necesidad de estudios de fragmentación en epífitas en cualquier lugar de los trópicos.

Estudios caso en conjuntos de epífitas vasculares han fallado en encontrar una respuesta común y general a la perturbación humana. En el capítulo 5, una hipótesis que pretende explicar y unir estos patrones conflictivos en la luz de un efecto no linear de clima local es propuesta y explorada: mientras los diversos conjuntos de bosques nublados, ligeramente estacionales cuentan con la integridad del dosel del bosque durante las sequías, esta dependencia puede ser menos pronunciada en bosques húmedos no estacionales o bosques secos distintamente. Esta hipótesis es apoyada por los resultados de un análisis de literatura.

ZUSAMMENFASSUNG

Während die Entwaldung der Tropen ungebremst voranschreitet, bleibt ungewiß, in welchem Maß anthropogene Habitate den Verlust an Diversität von Waldorganismen werden auffangen können. Fallstudien bezüglich Störungseffekten auf Epiphytengemeinschaften haben widersprüchliche Muster gezeigt, und die zugrundeliegenden Mechanismen sind ungeklärt. Eine gegenwärtig stark debattierte Frage ist beispielsweise der Einfluß von Wachstumsbedingungen gegenüber Verbreitungslimitierungen („dispersal constraints') auf die Fähigkeit von Epiphytenpopulationen, die Störung oder Modifizierung ihres Habitats zu überdauern. Die vorliegende Dissertation möchte zum Verständnis jener Prozesse beitragen, welche die Epiphytendiversität in anthropogenen Landschaften bestimmen. Dabei stützt sich diese Arbeit in erster Linie auf Gemeinschaften vaskulärer Epiphyten einzelstehender (isolierter) Bäume, welche sich als ein erstklassiges

Modellsystem anbieten. Da einzelstehende Bäume das kleinstmögliche Waldfragment darstellen, sind sie multiplen Waldrandeffekten ('edge effects') und potentiell eingeschränktem Diasporeneintrag ausgesetzt.

Die Feldarbeiten wurden von 2003–2006 an zwei Standorten in den Anden Ecuadors durchgeführt: (1) feuchter Bergwald (2200 mm Regen p.a.; 1800–2200 m NN) der Estación Científica San Francisco (ECSF) und angrenzende Matrixhabitate, auf der Andenostabdachung (Provinz Zamora-Chinchipe), und (2) perarider interandiner Bergwald (530 mm Regen p.a.; 2300 m NN) sowie angrenzende Brachen und Weideflächen im Schutzgebiet Bosque Protector Jerusalén (Provinz Pichincha).

Am humiden Standort (ECSF) wurde die Diversität vaskulärer Epiphyten auf Reliktbäumen untersucht, die seit ca. 10–30 Jahren auf Weidenflächen isoliert waren (Kapitel 2). Ziel dieser Studie war, den Wert von Reliktbäumen für die Erhaltung von Epiphytengemeinschaften zu dokumentieren. Trägerbäume wurden nach Johansson (1974) in fünf Zonen unterteilt und mit Einseil-Klettertechnik bestiegen. Es zeigte sich daß Abundanz und Diversität vaskulärer Epiphyten auf Reliktbäumen im Vergleich zu Waldbäumen signifikant niedriger waren. Eine Reihe von Pflanzenfamilien welche reich an xerophilen Taxa sind (Bromeliaceae, Orchidaceae, Piperaceae, Polypodiaceae) wiesen auf Reliktbäumen eine verhältnismäßig hohe Abundanz und Diversität auf. Andere, in erster Linie hygrophile Familien wie etwa Dryopteridaceae, Ericaceae, Grammitidaceae und Hymenophyllaceae waren dagegen stark unterrepräsentiert oder fehlten völlig. Die Verarmung war am stärksten an Stammbasen und in der äußeren Krone, und am schwächsten im Bereich der inneren Krone isolierter Trägerbäume ausgeprägt. Die Ergebnisse der Studie legen nahe, daß mikroklimatische Veränderungen für die beobachtete Verarmung von maßgeblicher Bedeutung sind.

Experimentelle Arbeiten am selben Standort lieferten Einblicke in die Mechanismen welche der beobachteten Verarmung zugrunde liegen. In einem Feldversuch (Kapitel 6) wurden Abundanz, Diversität und floristische Zusammensetzung von Epiphytenkeimlingen auf einzelstehenden Bäumen und in angrenzendem geschlossenem Wald untersucht. Dazu wurden sämtliche vaskulären Epiphyten von Stammbasen der häufigen Baumart *Piptocoma discolor* entfernt. Die Wiederbesiedlung durch Keimlinge wurden nach zwei Jahren erfaßt, und deren Überleben nach einem weiteren Jahr kontrolliert. Keimlingsdichte und -diversität (Anzahl von Familien bzw. Gattungen) je Untersuchungsfläche zeigten sich auf isolierten Bäumen signifikant niedriger als auf

Waldbäumen, ebenso wie die durch Rarefaction ermittelte Gesamtdiversität. Keimlingsgesellschaften auf Waldbäumen waren von hygrophilen Unterwuchsfarnen geprägt, wohingegen die Gesellschaften einzelstehender Bäume von xerophilen Taxa dominiert wurden. Die Besiedlungswahrscheinlichkeit war auf isolierten Bäumen in Waldnähe signifikant höher als auf Bäumen in größerer Entfernung zu Waldrändern, jedoch nicht auf Bäumen mit stärkerem Kronenschluß oder höherer Moosbedeckung. Die Keimlingssterblichkeit auf einzelstehenden Bäumen war für hygrophile Taxa signifikant höher als für xerophile Taxa. Diese Ergebnisse zeigen, daß limitierte Rekrutierung die Langzeitverarmung postjuveniler Epiphytengemeinschaften auf isolierten Bäumen zu erklären vermag. Die beobachtete stark veränderte Besiedlung isolierten Bäumen wird auf die Kombination von limitiertem Diasporeneintrag und mikroklimatischen Veränderungen zurückgeführt, welche durch kontinuierliche Messungen von Luftfeuchte und -temperatur dokumentiert wurden. Obgleich einzelstehende Bäume in vielerlei Hinsicht Schlüsselstrukturen für die Erhaltung der Diversität anthropogener Landschaften darstellen, zeigen die vorliegenden Ergebnisse, daß ihr Wert für die Erhaltung von Epiphytengemeinschaften begrenzt sein kann. Darüber hinaus könnten fortschreitende Habitatmodifizierung und globale Erwärmung durch ihren Einfluß auf abiotische Wachstumsbedingungen dazu führen, daß die Keimlingsetablierung zunehmend einen Engpaß im Lebenszyklus von Epiphyten darstellt.

In einem zweiten Feldversuch (Kapitel 6) wurden die Folgen von Waldstörung auf Gesellschaften fest etablierter, postjuveniler Epiphyten. Dazu wurden einzelne Gefäßpflanzen auf einzelstehenden Reliktbäumen (bis 5 m über dem Boden) in einer frischen Rodung über einen Zeitraum von drei aufeinanderfolgenden Jahren erfaßt. Durch Wind verursachte Baumstürze und Astbrüche führten zum Verlust von 24% der auf Reliktbäumen erfaßten Pflanzen, wohingegen sich die Verluste im Wald auf weniger als 1% beliefen. Nach drei Jahren war die Sterblichkeit vaskulärer Epiphyten auf den intakt gebliebenen Reliktbäumen mit 72% deutlich höher als im Wald (11%). Die höchste Sterblichkeitsrate auf Reliktbäumen (52%) wurde nach dem ersten Jahr verzeichnet, die niedrigste nach dem zweiten (20%). Die Mortalität war bei Pteridophyten und Dicotyledonae höher als bei Monocotyledonae (Araceae, Bromeliaceae, Orchidaceae). Jedoch zeigte sich die Sterblichkeit auf Reliktbäumen in allen größeren Pflanzenfamilien signifikant erhöht.

Die vorliegenden Ergebnisse liefern erstmalig einen experimentellen in situ-Nachweis dafür, daß eine verstärkte physische Exponierung die Vitalität gut etablierter vaskulärer Epiphyten stark beeinträchtigt. Veränderte abiotische Wachstumsbedingungen infolge verstärkter Exponierung könnten somit häufig einen einflußreicheren Faktor für die Diversität vaskulärer Epiphyten in anthropogen überformten Habitaten darstellen als Verbreitungslimitierungen. Zunehmender Trockenstreß ist dabei nur einer von mehreren solcher Randeffekte, welche die Vitalität von Epiphyten in gestörten Habitaten beeinträchtigen könnte.

Die beiden Versuche konnten aufzeigen, daß die starke Reduzierung der Epiphytendiversität, wie sie auf lokalen Reliktbäumen zu finden ist, sich sowohl auf reduzierte Rekrutierung als auch auf erhöhte Sterblichkeit bereits etablierter Pflanzen zurückführen läßt. Ferner wurde dokumentiert, daß veränderte Wachstumsbedingungen mittelfristig dramatische Folgen für Epiphytengesellschaften haben können.

Am Trockenwaldstandort Bosque Protector Jerusalén wurden Diversität und Abundanz von Epiphytengemeinschaften entlang eines Störungsgradienten untersucht (Kapitel 3 und 4). Dazu wurden epiphytische Moose und Gefäßpflanzen auf 100 Exemplaren der dominanten Baumart *Acacia macracantha* in fünf Habitattypen erfaßt: geschlossener Mischwald und reiner Akazienbestand (reifer Sekundärwald), Waldrand, halboffener junger Sekundärwald und einzelstehende Bäume auf alten Weideflächen. Trägerbäume wurden in vier Zonen unterteilt (cf. Johansson 1974), in welchen Diversität und Bedeckungsgrad von Gefäßpflanzen ermittelt wurden. Moose wurden dagegen nur in der inneren Krone erfaßt. Artenzahlen (pro Baum) waren am Waldrand und auf einzelstehenden Bäumen signifikant niedriger als in den beiden geschlossenen Waldtypen. Dies konnte auf eine starke Verarmung an Moosen zurückgeführt werden, wohingegen sich die Artenzahlen vaskulärer Epiphyten nicht signifikant zwischen Habitattypen unterschieden. Überraschenderweise zeigte sich die Epiphytengesellschaft von Waldrändern stärker verarmt als jene von jungem, halboffenem Sekundärwald, und zeigte darüber hinaus eine ähnlich große floristische Verwandtschaft zu einzelstehenden Bäumen wie zu geschlossenem Wald. Die Epiphytengesellschaften waren signifikant verschachtelt (‚nested'): in ihrer Artenzusammensetzung stellten Gesellschaften stärker gestörte Habitattypen lediglich Fragmente der Gesellschaften geschlossener Waldtypen dar, was einen graduellen Rückgang verfügbarer Nischen andeutet.

Diese Schlußfolgerung wird durch die bestehenden Abundanzmuster gestützt. Bromelien wuchsen am dichtesten in der mittleren Krone und zeigten sich somit xerophil und frühsukzessional, wohingegen Farne fast völlig auf die innere Krone beschränkt waren und sich somit schattenliebend und spätsukzessional zeigten. Während Bromelien ihre größte Abundanz auf einzelstehenden Bäumen erreichten, waren Farne auf geschlossene und halboffene Waldtypen beschränkt. Die Moosbedeckung war auf einzelstehenden Bäumen und am Waldrand substantiell niedriger als in geschlossenem Wald.

Die Entfernung zum Wald hatte keinen Einfluß auf Diversität oder Bedeckungsgrade von Epiphyten auf einzelstehenden Bäumen. Jedoch zeigte sich unter Einbezug aller Habitattypen eine enge Korrelation von Gesamtdiversität und Moosbedeckung mit dem Grad des Kronenschlusses. Diese Ergebnisse deuten stark darauf hin, daß mikroklimatische Veränderungen ein Schlüsselfaktor für die beobachtete Verarmung der Epiphytengemeinschaften mit zunehmendem Störungsgrad sind. Kryptogame Epiphyten scheinen sensible Indikatoren für Mikroklima und menschliche Störung zu sein. Ferner weisen die erzielten Ergebnisse darauf hin, daß die ‚similar gradient hypothesis' von McCune (1993) auch für tropische Trockenwaldgemeinschaften von Epiphyten gültig ist. Die starke Verarmung von Waldrändern unterstreicht den Bedarf an weiteren Fragmentierungsstudien zu tropischen Epiphyten.

Feldstudien konnten bislang keine übereinstimmenden Muster in der Reaktion vaskulärer Epiphytengemeinschaften auf menschliche Störung finden. In Kapitel 5 wird eine Hypothese postuliert und verfolgt, welche diese Diskrepanzen im Lichte einer nichtlinearen Abhängigkeit vom Lokalklima zu erklären sucht: während Epiphytengemeinschaften in feuchten, schwach saisonalen Wäldern stark auf die Integrität des Kronendachs angewiesen sind, könnte diese Abhängigkeit in asaisonal nassen und stark saisonalen (trockenen) Wäldern weniger stark ausgeprägt sein. Diese Hypothese wird von einer Literaturstudie gestützt.

Appendix 1. Host tree characteristics. Forest trees (FT) and isolated remnant trees (IRT).

No.	Host species	DBH [cm]	Height [m]	Elev. [m]	Years isolated	Species of epiphytes	Stands of epiphytes	Shannon H' log10	Simpson D
FT 1	Tapirira guianensis	35.5	17.3	1810	–	63	463	1.40	0.07
FT 2	Alzatea verticillata	32.1	16.3	2030	–	98	1209	1.28	0.12
FT 3	Alchornea pearcii	32.8	12.0	2230	–	90	2519	1.32	0.08
FT 4	Tabebuia chrysantha	38.2	19.5	1890	–	37	517	0.68	0.49
FT 5	Tabebuia chrysantha	33.7	20.1	1950	–	49	209	1.42	0.06
FT 6	Cedrela montana	32.8	17.2	1930	–	19	55	1.10	0.09
IRT 1	Tabebuia chrysantha	47.1	19.2	1860	30	22	872	0.42	0.65
IRT 2	Cedrela montana	39.1	13.7	1870	30	15	87	0.92	0.15
IRT 3	Juglans neotropica	27.1	13.2	1940	5	9	38	0.66	0.31
IRT 4	Juglans neotropica	35.0	17.2	2010	15	6	42	0.62	0.28
IRT 5	Tabebuia chrysantha	33.4	9.8	2060	12	5	9	0.62	0.19
IRT 6	Tabebuia chrysantha	50.3	16.4	2050	2	26	92	1.21	0.08
IRT 7	Tabebuia chrysantha	50.6	16.7	2050	12	10	22	0.90	0.12
IRT 8	Cedrela montana	42.3	14.5	2050	12	5	7	0.67	0.10
IRT 9	Tabebuia chrysantha	35.3	18.3	2130	14	12	64	0.67	0.30
IRT 10	Tabebuia chrysantha	40.7	22.2	2080	14	11	64	0.74	0.24
IRT 11	Tabebuia chrysantha	33.4	14.5	2130	14	7	51	0.63	0.27
IRT 12	Juglans neotropica	30.0	12.1	2040	14	2	3	0.28	0.33
IRT 13	Piptocoma discolor	35.0	8.0	2040	13	13	47	0.76	0.31
IRT 14	Heisteria sp. nov.	38.2	21.1	1840	30	20	119	1.10	0.11
IRT 15	Beilschmiedia costaricensis	35.7	18.8	1840	30	8	17	0.81	0.13

Appendix 2. Epiphyte species list.
Abbrevations of life forms include "AE" for accidental epiphytes, "E" for facultative and obligate holoepiphytes, "PH" for primary hemiepiphytes and "SH" for secondary hemiepiphytes.

	Life form	Stands FTs	Stands IRTs	Frequencies FTs	Frequencies IRTs
ALZATEACEAE					
Alzatea verticillata Ruiz & Pav.	PH	1	0	1	0
ARACEAE					
Anthurium dombeyanum Brogn.	E	0	2	0	1
Anthurium grubbii Croat	E	1	0	1	0
Anthurium scandens (Aubl.) Engl.	E	0	2	0	1
Anthurium cutucuense Madison vel aff.	SH	1	0	1	0
Philodendron ceronii Croat	SH	1	0	1	0
Philodendron sp. nov. 1	SH	1	0	1	0
Philodendron sp. nov. 2	SH	2	0	2	0
Stenospermation sp. nov.	E	1	0	1	0
ARALIACEAE					
Schefflera cf. *pentandra* (Ruiz & Pav.) Harms	PH	1	0	1	0
Schefflera sp.	PH	1	0	1	0
ASCLEPIADACEAE					
Matelea sp.	AE	1	0	1	0
ASTERACEAE					
Baccharis cf.	AE	1	0	1	0
Pentacalia cf. *moronensis* H.Rob. & Cuatrecas.	SH	3	0	1	0
BOMBACACEAE					
Spirotheca cf.	PH	1	0	1	0
BROMELIACEAE					
Guzmania coriostachya (Griseb.) Mez	E	3	0	1	0
Guzmania killipiani L.B.Sm.	E	3	0	1	0
Guzmania morreniana (Linden Hortus) Mez	E	3	0	1	0
Pitcairnea riparia Mez	SH	3	0	1	0
Racinaea dielsii (Harms) H.Luther	E	1	0	1	0
Racinaea euryelytra J.R.Grant	E	4	1	3	1
Racinaea monticola (Mez & Sodiro) M.A.Spencer & L.B.Sm.	E	21	1	2	1
Racinaea schumanniana (Wittm.) J.R.Grant	E	10	0	3	0
Racinaea tetrantha (Ruiz & Pav.) M.A.Spencer & L.B.Sm.	E	6	0	1	0
Racinaea undulifolia (Mez) H.Luther	E	9	0	2	0
Tillandsia barbeyana Wittm.	E	77	45	5	8
Tillandsia barthlottii Rauh	E	91	26	3	8
Tillandsia biflora Ruiz & Pav.	E	4	11	3	4
Tillandsia complanata Benth.	E	24	121	5	14
Tillandsia confinis var. *caudata* L.B.Sm.	E	2	0	1	0
Tillandsia fendleri Griseb.	E	3	5	3	4
Tillandsia laminata L.B.Sm.	E	28	0	2	0
Tillandsia naundorffiae Rauh & Barthlott	E	98	10	6	5
Tillandsia stenoura Harms	E	0	1	0	1
Tillandsia tovariensis Mez	E	30	102	5	12
Vriesea appendiculata (L.B.Sm.) L.B.Sm.	E	73	47	4	11
Vriesea fragrans (André) L.B.Sm.	E	5	0	1	0
Vriesea incurva (Griseb.) Read	E	0	3	0	3
Vriesea lutherii J.M.Manzanares & W.Till	E	2	0	1	0
Vriesea tequendamae (André) L.B.Sm.	E	2	1	1	1

Appendix 2 (continued).

	Life form	Stands FTs	Stands IRTs	Frequencies FTs	Frequencies IRTs
CACTACEAE					
Rhipsalis riocampanensis J.E.Madsen & Z.Aguirre	E	1	0	1	0
CLUSIACEAE					
Clusia cf. *alata* Triana & Planch.	PH	1	0	1	0
Clusia cf. *ducuoides* Engl.	PH	1	0	1	0
Clusia sp.	PH	2	0	2	0
CUNONIACEAE					
Weinmannia pubescens Kunth	AE	2	0	1	0
CYCLANTHACEAE					
Asplundia sp.	SH	5	0	2	0
DRYOPTERIDACEAE					
Elaphoglossum latifolium (Sw.) J.Sm.	E	3	0	1	0
Elaphoglossum oleandropsis (Sodiro) Christ	E	94	1	4	1
Elaphoglossum cf. *craspedotum* Copel.	AE	1	0	1	0
Elaphoglossum cf. *cuspidatum* (Willd.) T.Moore	E	1	0	1	0
Elaphoglossum cf. *muscosum* (Sw.) T.Moore	E	4	0	3	0
Elaphoglossum cf. *pachyphyllum* (Kunze) C.Chr.	E	1	0	1	0
Elaphoglossum cf. *rimbachii* (Sodiro) H.Christ	E	79	0	4	0
Elaphoglossum sp. 1	E	6	0	2	0
Elaphoglossum sp. 2	E	1	0	1	0
ERICACEAE					
Cavendishia isernii Sleumer vel aff.	SH	1	0	1	0
Cavendishia cf.	E	1	0	1	0
Ceratostema loranthifolium Benth. vel aff.	E	1	0	1	0
Disterigma pentandrum S.F.Blake	E	21	0	2	0
Macleania mollis A.C.Sm.	E	26	0	2	0
Macleania hirtifolia (Benth.) A.C.Sm. vel aff.	E	1	0	1	0
Oreanthus cf. *hypogaeus* (A.C.Sm.) Luteyn	SH	1	0	1	0
Cavendishia cf. *bracteata* (Ruiz & Pav. ex. J.St.-Hil.) Hoerold	SH	1	0	1	0
Psammisia sp.	E	1	0	1	0
Semiramisia speciosa (Benth.) Klotzsch	E	1	0	1	0
Sphyrospermum cordifolium Benth.	E	11	1	2	1
Sphyrospermum sp.	E	17	0	3	0
Thibaudia vel aff.	E	1	0	1	0
GESNERIACEAE					
Columnea sp.	SH	1	0	1	0
GRAMMITIDACEAE					
Ceradenia melanopus (Grev. & Hook.) Bishop	E	4	0	1	0
Cochlidium serrulatim (Sw.) L.E.Bishop vel aff.	E	1	0	1	0
Enterosora sp.	E	1	0	1	0
Grammitis paramicola L.E.Bishop	E	11	0	1	0
Lellingeria subsessilis (Baker) A.R.Sm. & R.C.Moran	E	1	0	1	0
Melpomene anfractuosa (Kl.) A.R.Sm. & R.C.Moran	E	1	0	1	0
Melpomene cf. *pilosissima* (Mart.& Gal.) A.R.Sm.& R.C.Moran	E	1	0	1	0
Melpomene firma (J.Sm.) A.R.Sm.& R.C.Moran	E	20	0	1	0
Melpomene flabelliformis (Poiret) A.R.Sm. & R.C.Moran	E	488	0	4	0
Melpomene xiphopteroides (Liebm.) A.R.Sm. & R.C.Moran	E	29	2	2	1
Terpsichore pichinchae (Sodiro) A.R.Sm.	E	1	0	1	0

Appendix 2 (continued).

	Life form	Stands FTs	IRTs	Frequencies FTs	IRTs
Terpsichore sp.	E	5	0	2	0
Zygophlebia matthewsii (Kunze ex. Mett) L.E.Bishop	E	2	0	1	0
HYDRANGEACEAE					
Hydrangea sp.	SH	2	0	1	0
HYMENOPHYLLACEAE					
Hymenophyllum fucoides var. *fucoides* (Sw.) Sw.	E	11	0	2	0
Hymenophyllum lindenii Hooker	E	4	0	1	0
Hymenophyllum myriocarpum Hook.	E	13	0	2	0
Hymenophyllum plumosum Kaulf.	E	1	0	1	0
Hymenophyllum ruizianum (Kl.) Kunze	E	1	0	1	0
Hymenophyllum trichophyllum Kunth	E	5	0	1	0
Hymenophyllum undulatum (Sw.) Sw.	E	36	0	2	0
Hymenophyllum sp.	E	1	0	1	0
Trichomanes hymenoides Hedw.	E	1	0	1	0
Trichomanes lucens Sw.	E	1	0	1	0
LENTIBULARIACEAE					
Utricularia jamesoniana Oliver	E	592	0	3	0
MARCGRAVIACEAE					
Marcgravia sp. nov.	SH	2	0	1	0
Ruyschia sp. nov.	SH	1	0	1	0
MELASTOMATACEAE					
Blakea vel aff. 1	E	2	0	1	0
Blakea vel aff. 2	SH	2	0	1	0
Blakea subpanduriforme Cotton & Matezki	SH	1	0	1	0
Blakea sp.	SH	1	0	1	0
Clidemia sp.	SH	1	0	1	0
MORACEAE					
Ficus krukovii Standl.	PH	1	0	1	0
Ficus sp. 1	PH	0	2	0	2
Ficus sp. 2	PH	0	1	0	1
ORCHIDACEAE					
Barbosella cucullata (Lindl.) Schltr.	E	3	0	2	0
Cochlioda rosea (Lindl.) Benth.	E	0	1	0	1
Cranichis sp.	E	24	0	1	0
Cryptocentrum cf. *lehmanni* (Rchb.f.) Garay	E	183	0	2	0
Cyrtochilum sp.	E	36	0	1	0
Dryadella werneri Luer	E	396	735	2	2
Elleanthus bifarius Garay	E	4	0	2	0
Elleanthus blatteus Garay	E	1	0	1	0
Elleanthus robustus (Rchb.f.) Rchb.f.	E	7	0	2	0
Epidendrum gracilimum Rchb.f. & Warsz.	E	343	0	2	0
Epidendrum loxense F.Lehm. & Kraenzl.	E	1	0	1	0
Epidendrum repens Cogn.	E	4	0	1	0
Epidendrum sophronitoides F.Lehm. & Kraenzl.	E	5	0	2	0
Epidendrum stangeatum Rchb.f.	E	3	17	1	4
Epidendrum cf. *zosterifolium* F.Lehm. & Kraenzl.	E	19	108	3	10
Epidendrum cf. *coryophorum* (Kunth.) Rchb.f.	E	80	13	3	3
Epidendrum mancum Lindl.	E	2	0	1	0

	Life form	Stands FTs	IRTs	Frequencies FTs	IRTs
Epidendrum sp. 1	E	6	0	1	0
Epidendrum sp. 2	E	0	1	0	1

Appendix 2 (continued).

	Life form	Stands FTs	IRTs	Frequencies FTs	IRTs
Epidendrum sp. 3	E	1	0	1	0
Epidendrum sp. 4	E	0	5	0	4
Epidendrum cf.	E	0	2	0	1
Fernandezia subbiflora Ruiz & Pav.	E	42	0	1	0
Kefersteinia sp.	E	0	3	0	1
Lepanthes wageneri Rchb.f.	E	71	6	4	1
Lepanthopsis acuminata Ames	E	9	0	1	0
Lepanthopsis floripecten (Rchb.f.) Ames	E	134	0	1	0
Lycaste ciliata (Ruiz & Pav.) Lindl. & Rchb.f.	E	1	1	1	1
Masdevallia bangii Schlchtr.	E	1	0	1	0
Masdevallia bicolor Poepp. & Endl.	E	2	0	2	0
Masdevallia persicina Luer	E	3	0	1	0
Maxillaria acuminata Lindl.	E	24	0	5	0
Maxillaria aggregata Enders	E	30	0	3	0
Maxillaria brevifolia (Lindl.) Rchb.f.	E	8	0	2	0
Maxillaria cryptobulbon Carnevali & A.T.Atwood	E	2	0	2	0
Maxillaria imbricata Barb. Rodriguez	E	25	0	2	0
Maxillaria jenishiana (Rchb.f.) C.Schweinf.	E	4	0	2	0
Maxillaria mapiriensis (Kraenzl.) L.O.Williams	E	43	0	3	0
Maxillaria notylioglossa Rchb.f.	E	5	0	1	0
Maxillaria ochroleuca Lodd. ex. Lindl.	E	6	0	1	0
Maxillaria polyphylla Rchb.f.	E	10	0	2	0
Maxillaria rufescens Lindl.	E	19	4	3	1
Maxillaria stenophylla Rchb.f.	E	3	0	1	0
Maxillaria cf. *pulla* Linden & Rchb.f.	E	14	0	1	0
Maxillaria cf. *xantholeuca* Schlchtr.	E	5	0	1	0
Maxillaria calantha Schlchtr. vel aff.	E	5	0	2	0
Maxillaria notylioglossa Rchb.f. vel aff.	E	1	0	1	0
Maxillaria sp.	E	1	0	1	0
Myoxanthus affinis (Lindl.) Luer	E	1	0	1	0
Myoxanthus ceratothallis Luer	E	1	0	1	0
Myoxanthus uxorius Luer	E	3	0	1	0
Myoxanthus sp.	E	1	0	1	0
Odontoglossum sp.	E	0	10	0	1
Oncidium sp. 1	E	25	3	1	1
Oncidium sp. 2	E	114	0	2	0
Pityphyllum pinoides Sweet	E	88	0	2	0
Pleurothallis crocodiliceps Rchb.f.	E	4	0	2	0
Pleurothallis decurrens Poepp. & Endl.	E	0	5	0	2
Pleurothallis galeata Lindl.	E	8	0	1	0
Pleurothallis lilijae Foldats	E	1	0	1	0
Pleurothallis palateensis Luer	E	3	0	1	0
Pleurothallis peroniocephala Luer	E	3	0	2	0
Pleurothallis rabei Foldats	E	21	3	3	1
Pleurothallis rubens Lindl.	E	26	0	4	0
Pleurothallis talpinaria Rchb.f.	E	1	0	1	0
Pleurothallis xanthochlora Rchb.f.	E	4	0	1	0
Pleurothallis cf. *bivalvis* Lindl.	E	18	0	6	0

Pleurothallis cf. *erinacea* Rchb.f.		E	0	1	0	1
Pleurothallis sp. 1		E	1	0	1	0

Appendix 2 (continued).

	Life form	Stands FTs	Stands IRTs	Frequencies FTs	Frequencies IRTs
Pleurothallis sp. 2	E	27	0	3	0
Pleurothallis sp. 3	E	5	0	1	0
Pleurothallis sp. 4	E	1	0	1	0
Pleurothallis sp. 5	E	1	0	1	0
Pleurothallis sp. 6	E	2	0	1	0
Pleurothallis sp. 7	E	1	0	1	0
Pleurothallis sp. 8	E	1	0	1	0
Pleurothallis cf. sp. 1	E	1	0	1	0
Pleurothallis cf. sp. 2	E	0	1	0	1
Pleurothallis cf. sp. 3	E	9	0	1	0
Pleurothallis cf. sp. 4	E	13	3	1	1
Pleurothallis cf. sp. 5	E	1	0	1	0
Pleurothallis cf. sp. 6	E	1	0	1	0
Pleurothallis cf. sp. 7	E	0	1	0	1
Polystachya stenophylla Schltr.	E	17	0	3	0
Prosthechea grammatoglossa (Rchb.f.) W.E.Higgins	E	4	43	2	5
Prosthechea pulchra Dodson & Higgins	E	1	0	1	0
Prosthechea vespa (Vell.) W.E.Higgins	E	11	0	2	0
Prosthechea cf. *hartwegii* (Lindl.) W.E.Higgins	E	11	0	1	0
Psilochilus mollis Garay	AE	1	0	1	0
Scaphyglottis bicornis (Lindl.) Garay	E	24	0	1	0
Scaphyglottis punctulata (Rchb.f.) C.Schweinf.	E	27	1	3	1
Sobralia candida Poepp. & Endl.	E	1	0	1	0
Sobralia cf. *croecea* Poepp. & Endl.	E	1	0	1	0
Stelis floriani Luer	E	1	0	1	0
Stelis sp. 1	E	1	0	1	0
Stelis sp. 2	E	16	2	2	1
Stelis sp. 3	E	1	0	1	0
Stelis sp. 4	E	4	0	2	0
Stelis sp. 5	E	1	0	1	0
Stelis sp. 6	E	12	3	2	2
Stelis sp. 7	E	0	1	0	1
Stelis sp. 8	E	0	1	0	1
Stelis sp. 9	E	0	1	0	1
Stelis sp. 10	E	0	7	0	1
Stelis sp. 11	E	4	0	1	0
Stelis sp. 12	E	8	6	1	1
Stelis sp. 13	E	1	0	1	0
Stelis sp. 14	E	1	0	1	0
Stelis sp. 15	E	252	0	1	0
Stelis sp. 16	E	17	0	1	0
Stelis sp. 17	E	350	0	1	0
Stelis sp. 18	E	2	0	1	0
Stelis cf. 1	E	8	0	2	0
Stelis cf. 2	E	2	0	1	0
Trichopilia fragrans (Lindl.) Rchb.f.	E	0	14	0	1
Trichosalpinx berlineri Luer	E	13	2	2	2
Trichosalpinx intricata (Lindl.) Luer	E	1	0	1	0

	Life form	Stands		Frequencies	
		FTs	IRTs	FTs	IRTs
Trichosalpinx robledorum Luer & Escobar	E	1	0	1	0
Trichosalpinx werneri Luer	E	17	0	1	0

Appendix 2 (continued).

	Life form	Stands		Frequencies	
		FTs	IRTs	FTs	IRTs
Trichosalpinx sp.	E	19	0	2	0
PIPERACEAE					
Peperomia ceroderma Yunck.	E	0	1	0	1
Peperomia ciliaris C.DC.	E	1	0	1	0
Peperomia hartwegiana Miq.	E	6	11	3	4
Peperomia tetraphylla (G.Forst.) Hook. & Arn.	E	11	5	3	2
Peperomia tovariana C.DC.	E	10	0	2	0
Peperomia vulcanicula vel aff.	E	5	3	2	1
Peperomia sp. 1	E	1	1	1	1
Peperomia sp. 2	E	0	1	0	1
Peperomia sp. 3	E	1	2	1	1
Peperomia sp. 4	E	2	0	1	0
Peperomia sp. 5	E	14	0	1	0
Piper sp.	SH	1	0	1	0
POLYPODIACEAE					
Campyloneuron amphostemon (Kunze ex. Klotzsch) Fée	E	1	1	2	1
Campyloneuron angustifolium (Sw.) Fée	E	0	5	0	3
Niphidium amocarpos (Kunze) Lellinger	E	0	1	0	1
Niphidium cf. *crassifolium* (L.) Lellinger	E	1	0	1	0
Pecluma eurybasis (C.Chr.) M.G.Price	E	2	0	1	0
Pleopeltis macrocarpa (Bory ex.Willd.) Kaulf.	E	21	55	3	7
Pleopeltis percussa (Cav.) Hook & Grev.	E	0	1	0	1
Polypodium fraxinifolium Jacq.	E	1	0	1	0
Polypodium levigatum Cav.	E	1	0	1	0
Polypodium remotum Desv.	E	11	48	4	4
Polypodium sessilifolium Desv.	E	3	4	1	2
Polypodium loriceum L.	E	1	0	1	0
RUBIACEAE					
Palicourea cf.	AE	1	0	1	0
SOLANACEAE					
Solanum sp.	SH	0	2	0	2
Trianea sp.	E	0	4	0	2
URTICACEAE					
Pilea sp.	AE	2	0	1	0
VITTARIACEAE					
Vittaria stipitata Kunze	E	26	0	3	0

Appendix 3. Frequency and abundance of vascular holoepiphytes and accidental epiphytes in five habitats (n = 20 trees).

Taxon	Coll. n°	N	Frequency/abundance[a]					Growth/ Life form
			MF	AF	FE	SW	IT	
Holoepiphytes								
Bromeliaceae								
Racinaea fraseri (Baker) M.A. Spencer & L.B. Sm.	fw836	14/40	3/3	2/3	–/–	6/12	3/22	tank-forming
Tillandsia incarnata Kunth	fw845	100*	20	20	20	20	20	atmospheric
Tillandsia lajensis André	fw833	5/7	1/2	2/2	–/–	–/–	2/3	tank-forming
Tillandsia recurvata (L.) L.	fw1977	100*	20	20	20	20	20	atmospheric
Tillandsia usneoides (L.) L.	fw1978	99*	20	20	20	20	19	atmospheric
Polypodiaceae								
Pleopeltis macrocarpa (Bory ex. Willd.) Kaulf.	fw814	4/4	2/2	1/1	–/–	1/1	–/–	desicc.-tolerant (?)
Pleopeltis murorum (Hook.) A.R. Sm., comb. ined.	fw851	2/2	1/1	1/1	–/–	–/–	–/–	desicc.-tolerant (?)
Pleopeltis thyssanolepis (A. Braun ex Klotzsch) E.G. Andrews & Wigham	fw815	2/2	2/2	–/–	–/–	–/–	–/–	desicc.-tolerant (?)
Accidental epiphytes								
Solanaceae								
Capsicum rhomboideum (Dunal) Kuntze	fw816	15/23	2/3	–/–	3/3	5/6	5/11	subwoody
Solanum sp.	fw835	3/3	1/1	–/–	–/–	1/1	1/1	subwoody
Verbenaceae								
Lantana camara L.	–	1/2	–/–	–/–	–/–	1/2	–/–	woody
Cactaceae								
Cleistocactus sepia (Kunth) F.A.C. Weber	fw848	3/3	1/1	–/–	–/–	1/1	1/1	subwoody
Poaceae								
Poaceae indet.	fw829	1/1	–/–	–/–	–/–	–/–	1/1	herbaceous
Asteraceae								
Bidens cf. *cynapifolia* Kunth	fw817	1/1	–/–	–/–	–/–	–/–	1/1	herbaceous
Polygonaceae								
Polygonum sp.	fw831	1/2	–/–	1/2	–/–	–/–	–/–	herbaceous
Crassulaceae								
Bryophyllum pinnatum (Lam.) Oken	fw839	1/1	–/–	–/–	–/–	1/1	–/–	herbaceous

[a] habitat types: MF = mature mixed forest, AF = mature acacia forest, FE = mature forest edge, SW = semi-closed woodland, IT = isolated trees.
* no individual counts available

Appendix 4. Spatial distribution (relative frequencies) of vascular holoepiphytes in five habitats ($n = 20$ trees).

Taxon	MF					AF					FE					SW					IT				
	JZ1	JZ3	JZ4	JZ5		JZ1	JZ3	JZ4	JZ5		JZ1	JZ3	JZ4	JZ5		JZ1	JZ3	JZ4	JZ5		JZ1	JZ3	JZ4	JZ5	
Bromeliaceae																									
Racinaea fraseri (Baker) M.A. Spencer and L.B. Sm.	–	–	0.05	0.10		–	–	0.05	0.05		–	–	–	–		–	–	–	0.05	0.30	–	–	0.05	0.10	0.15
Tillandsia incarnata Kunth	0.10	1.00	1.00	1.00		0.05	1.00	1.00	1.00		0.15	1.00	1.00	1.00		–	0.40	1.00	1.00	1.00	0.05	0.95	0.95	1.00	
Tillandsia lajensis André	–	0.05	–	0.05		–	–	–	0.10		–	–	–	–		–	–	–	–	–	–	–	–	0.05	0.10
Tillandsia recurvata (L.) L.	0.15	1.00	1.00	1.00		0.05	1.00	1.00	1.00		0.25	1.00	1.00	1.00		–	0.50	1.00	1.00	1.00	0.80	1.00	1.00	1.00	
Tillandsia usneoides (L.) L.	–	0.80	1.00	1.00		–	0.05	0.95	1.00	1.00	–	0.05	0.85	1.00	1.00	–	0.30	1.00	1.00	1.00	0.05	0.85	0.95	0.95	
Polypodiaceae																									
Pleopeltis macrocarpa (Bory ex. Willd.) Kaulf.	–	0.10	–	–		–	–	0.05	–		–	–	–	–		–	–	0.05	–	–	–	–	–	–	
Pleopeltis murorum (Hook.) A.R. Sm., comb. ined.	–	0.05	–	–		–	–	0.05	–		–	–	–	–		–	–	–	–	–	–	–	–	–	
Pleopeltis thyssanolepis (A. Braun ex Klotzsch) E.G. Andrews and Wigham	–	0.10	–	–		–	–	–	–		–	–	–	–		–	–	–	–	–	–	–	–	–	

Johansson-zones (JZ; modified after Johansson 1974): JZ1 = trunk, JZ3 = inner crown, JZ4 = middle crown, JZ5 = outer crown.

Appendix 5. Annual rates of mortality (mort.) at genus level on isolated trees (IRTs) and control trees (forest).

Taxon	2003-2004 Forest n	2003-2004 Forest mort.	2003-2004 IRTs n	2003-2004 IRTs mort.	2004-2005 Forest n	2004-2005 Forest mort.	2004-2005 IRTs n	2004-2005 IRTs mort.	2005-2006 Forest n	2005-2006 Forest mort.	2005-2006 IRTs n	2005-2006 IRTs mort.	2003-2006 Forest mort.	2003-2006 IRTs mort.
Araceae														
Anthurium	18	0.0	6	16.7	18	0.0	5	0.0	18	0.0	5	0.0	0.0	16.7
Philodendron	3	0.0	7	14.3	3	0.0	6	0.0	3	33.3	5	40.0	33.3	48.6
Stenospermation	1	0.0	–	–	1	0.0	–	–	1	100	–	–	100	–
Aspleniaceae														
Asplenium	43	4.7	27	92.6	41	4.9	2	50.0	39	17.9	1	0.0	25.6	96.3
Asteraceae														
Pentacalia	–	–	1	100	–	–	0	–	–	–	–	–	–	100
Blechnaceae														
Blechnum	1	0.0	1	100	1	0.0	0	–	1	100	–	–	100	100
Bromeliaceae														
Guzmania	222	2.7	90	31.1	214	3.3	62	12.9	206	5.3	51	27.5	10.9	56.5
Pitcairnia	9	0.0	7	14.3	9	0.0	6	16.7	9	0.0	5	40.0	0.0	57.1
Racinaea	–	–	1	0.0	–	–	1	0.0	–	–	1	0.0	–	0.0
Tillandsia	1	0.0	1	0.0	1	0.0	1	0.0	1	0.0	1	0.0	0.0	0.0
Clusiaceae														
Clusia	1	0.0	4	50.0	1	0.0	2	0.0	1	0.0	2	100	0.0	100
Cyclanthaceae														
Sphaeradenia	–	–	1	100	–	–	0	–	–	–	–	–	–	100
Davalliaceae														
Nephrolepis	1	0.0	–	–	1	0.0	–	–	1	0.0	–	–	0.0	–
Dryopteridaceae														
Elaphoglossum	113	1.8	39	61.5	111	1.8	15	46.7	109	1.8	7	0.0	5.3	79.5
Ericaceae														
Cavendishia	–	–	1	100	–	–	–	–	–	–	–	–	–	100
Disterigma	1	0.0	–	–	1	0.0	–	–	1	0.0	–	–	0.0	–
Semiramisia	3	0.0	2	100	3	0.0	–	–	3	0.0	–	–	0.0	100
Sphyrospermum	2	0.0	2	50.0	2	0.0	1	0.0	2	0.0	1	0.0	0.0	50.0
Gesneriaceae														
Alloplectus	1	0.0	–	–	1	0.0	–	–	1	0.0	–	–	0.0	–
Columnea	1	0.0	1	0.0	1	0.0	1	0.0	1	0.0	1	100.0	0.0	100

Appendix 5 (continued).

Taxa	2003-2004 Forest n	2003-2004 Forest mort.	2003-2004 IRTs n	2003-2004 IRTs mort.	2004-2005 Forest n	2004-2005 Forest mort.	2004-2005 IRTs n	2004-2005 IRTs mort.	2005-2006 Forest n	2005-2006 Forest mort.	2005-2006 IRTs n	2005-2006 IRTs mort.	2003-2006 Forest mort.	2003-2006 IRTs mort.
Grammitidaceae														
Cochlidium	18	0.0	7	57.1	18	0.0	3	33.3	18	11.1	2	50.0	11.1	85.7
Grammitis	2	0.0	–	–	2	0.0	–	–	2	0.0	–	–	0.0	–
Lellingeria	2	0.0	19	52.6	2	0.0	9	33.3	2	0.0	6	16.7	0.0	73.7
Melpomene	8	0.0	1	0.0	8	25.0	1	100	6	50.0	–	–	62.5	100
Micropolypodium	4	0.0	–	–	4	0.0	–	–	4	0.0	–	–	0.0	–
Terpsichore	2	0.0	–	–	2	0.0	–	–	2	50.0	–	–	50.0	–
Hymenophyllaceae														
Hymenophyllum	58	0.0	63	73.0	58	0.0	17	35.3	58	3.4	11	45.5	3.4	90.5
Trichomanes	3	0.0	–	–	3	0.0	–	–	3	0.0	–	–	0.0	–
Lycopodiaceae														
Huperzia	7	0.0	3	66.7	7	0.0	1	100	7	0.0	–	–	0.0	100
Orchidaceae														
Brachycladium	–	–	41	53.7	–	–	19	36.8	–	–	11	63.6	–	89.4
Cryptocentrum	4	0.0	–	–	4	0.0	–	–	4	0.0	–	–	0.0	–
Cyrtochilum	–	–	1	0.0	–	–	1	0.0	–	–	1	0.0	–	0.0
Dichaea	4	0.0	1	0.0	4	0.0	1	0.0	4	0.0	1	0.0	0.0	0.0
Elleanthus	12	0.0	4	100	12	8.3	–	–	11	0.0	–	–	8.3	100
Epidendrum	14	7.1	2	0.0	13	0.0	2	50.0	13	0.0	1	0.0	7.1	50.0
Galleotia	2	0.0	–	–	2	0.0	–	–	2	0.0	–	–	0.0	–
Lepanthes	7	0.0	1	100	6	0.0	0	–	6	33.3	0	–	33.3	100
Lycaste	3	0.0	–	–	3	0.0	–	–	3	0.0	–	–	0.0	–
Masdevallia	2	0.0	1	0.0	2	0.0	1	0.0	2	0.0	1	0.0	0.0	0.0
Maxillaria	21	14.3	3	0.0	18	11.1	3	0.0	16	0.0	3	0.0	23.8	0.0
Myoxanthus	1	0.0	1	0.0	1	0.0	1	0.0	1	0.0	1	0.0	0.0	0.0
Oncidium	13	15.4	2	0.0	11	0.0	2	0.0	11	9.1	2	0.0	23.1	0.0
Pleurothallis	54	1.9	29	31.0	53	0.0	20	5.0	53	7.5	19	5.3	9.3	37.9
Stelis	61	0.0	19	36.8	61	3.3	12	0.0	58	8.6	12	16.7	11.6	47.4
Trichosalpinx	3	0.0	2	50.0	3	0.0	1	0.0	3	0.0	1	0.0	0.0	50.0

Appendix 5 (continued).

Taxa	2003-2004 Forest		2003-2004 IRTs		2004-2005 Forest		2004-2005 IRTs		2005-2006 Forest		2005-2006 IRTs		2003-2006 Forest	2003-2006 IRTs
	n	mort.	n	mort.	n	mort.	n	mort.	n	mort.	n	mort.	mort.	mort.
Piperaceae														
Peperomia	47	2.1	14	50.0	46	0.0	7	14.3	45	11.1	5	60.0	13.0	82.9
Polypodiaceae														
Campyloneurum	2	0.0	2	50.0	2	0.0	1	0.0	2	0.0	1	0.0	0.0	50.0
Niphidium	1	0.0	–	–	1	0.0	–	–	1	0.0	–	–	0.0	–
Pleopeltis	5	0.0	–	–	5	0.0	–	–	5	0.0	–	–	0.0	–
Polypodium	17	5.9	4	50.0	16	0.0	2	50.0	16	6.3	1	0.0	11.8	75.0
Pecluma	3	0.0	5	100	3	33.3	0	–	2	50.0	0	–	66.7	100
Urticaceae														
Pilea	1	0.0	2	100	1	0.0	0	–	1	100	0	–	100	100
Vittariaceae														
Polytaenium	6	0.0	8	100	6	0.0	0	–	6	0.0	0	–	0.0	100
Radiovittaria	39	0.0	20	50.0	39	2.6	10	40.0	38	5.3	5	40.0	7.7	82.0

Apendix 6. Relative annual increment (yr_x/yr_{x-1}) in leaf number and maximum leaf length on isolated remnant trees (IRTs) and control trees (forest). P-values yielded by Mann-Whitney U-test (values < 0.05 in bold letters).

		Leaf number					Maximum leaf length				
		IRTs		Forest			IRTs		Forest		
Taxon	Yr	n	Median	n	Median	P	n	Median	n	Median	P
Araceae	1	11	0.63	22	1.00	>0.1	9	0.68	22	0.99	**<0.01**
	2	11	1.00	22	1.00	>0.2	10	0.56	22	1.00	**<0.005**
	3	8	1.00	19	1.00	>0.2	8	0.97	20	1.00	>0.2
Aspleniaceae	1	2	0.73	38	1.00	>0.5	2	0.75	36	0.99	>0.5
	2	1	0.80	36	1.00	–	1	1.03	35	0.98	–
	3	1	1.00	28	0.96	–	1	1.00	28	1.04	–
Blechnaceae	1	–	–	1	1.20	–	–	–	1	0.90	–
	2	–	–	1	1.00	–	–	–	1	1.06	–
	3	–	–	–	–	–	–	–	–	–	–
Bromeliaceae	1	70	1.00	222	1.12	**<0.005**	71	0.66	224	1.05	**<0.0001**
	2	59	0.90	216	1.11	**<0.0001**	59	0.92	215	1.02	**<0.005**
	3	41	1.00	204	1.07	**<0.05**	41	0.92	204	1.01	**<0.0005**
Clusiaceae	1	1	1.09	1	0.54	–	1	0.18	1	1.05	–
	2	–	–	1	1.29	–	–	–	1	0.99	–
	3	–	–	1	1.11	–	–	–	1	0.92	–
Davalliaceae	1	–	–	1	0.33	–	–	–	1	0.41	–
	2	–	–	1	2.50	–	–	–	1	2.01	–
	3	–	–	1	1.40	–	–	–	1	0.99	–
Dryopteridaceae	1	16	0.92	111	1.00	>0.2	16	0.23	110	0.99	**<0.0001**
	2	8	1.48	109	1.00	**<0.005**	7	0.96	108	1.02	>0.5
	3	7	0.75	107	0.88	>0.2	6	0.75	107	0.97	**<0.005**
Ericaceae	1	1	1.37	6	0.86	–	1	1.03	6	1.04	–
	2	1	1.24	6	1.11	–	1	1.00	6	1.01	–
	3	1	0.72	6	0.99	–	1	0.97	6	0.99	–
Gesneriaceae	1	1	1.30	2	0.87	–	1	0.46	2	1.08	–
	2	1	0.69	2	0.76	–	1	0.57	2	1.27	–
	3	–	–	2	1.25	–	–	–	2	0.80	–
Grammitidaceae	1	13	0.50	35	1.00	**<0.0001**	13	0.43	33	1.00	**<0.0001**
	2	8	0.65	34	1.18	**<0.05**	8	0.61	32	1.00	>0.1
	3	6	0.69	27	0.60	>0.2	6	0.64	28	0.58	>0.5
Hymenophyllaceae	1	16	0.47	40	1.00	**<0.0001**	16	0.71	41	1.01	**<0.0005**
	2	10	1.00	26	1.12	>0.1	10	0.49	35	1.01	**<0.0005**
	3	6	1.29	5	0.70	>0.05	6	0.94	36	0.95	>0.5
Lycopodiaceae	1	1	0.92	7	0.99	–	–	–	4	1.07	–
	2	–	–	7	1.00	–	–	–	4	1.10	–
	3	–	–	7	0.97	–	–	–	7	1.11	–
Orchidaceae	1	64	0.86	191	1.00	**<0.001**	40	0.89	185	1.00	**<0.0001**
	2	54	1.18	185	1.00	<0.1	40	0.89	181	1.02	**<0.0001**
	3	41	1.13	173	1.00	**<0.05**	36	0.95	169	0.99	>0.1
Piperaceae	1	7	1.00	46	1.13	>0.1	7	0.78	46	1.00	**<0.05**
	2	6	0.77	45	1.04	>0.5	6	0.70	45	1.00	**<0.0001**
	3	2	0.96	39	1.00	–	2	1.06	39	1.00	–

Apendix 6 (continued).

Taxon	Yr	Leaf number					Maximum leaf length				
		IRTs		Forest			IRTs		Forest		
		n	Median	n	Median	P	n	Median	n	Median	P
Polypodiaceae	1	3	0.81	27	1.00	–	2	0.32	26	0.97	–
	2	2	0.63	26	1.00	–	2	1.13	25	1.08	–
	3	2	0.94	23	1.00	–	2	0.89	23	0.96	–
Vittariaceae	1	8	0.85	44	1.00	>0.1	8	0.32	36	1.00	**<0.0001**
	2	3	1.00	43	1.17	–	3	0.84	35	1.00	–
	3	1	1.29	41	0.73	–	1	1.09	34	0.95	–
Monocots	1	145	0.91	435	1.00	**<0.0001**	120	0.71	431	1.01	**<0.0001**
	2	124	1.00	423	1.06	**<0.05**	109	0.91	418	1.02	**<0.0001**
	3	90	1.00	396	1.00	>0.5	85	0.95	393	1.00	**<0.0001**
Pteridophytes	1	59	0.60	304	1.00	**<0.0001**	57	0.37	288	1.00	**<0.0001**
	2	32	1.00	283	1.00	>0.2	31	0.78	276	1.02	**<0.0005**
	3	23	0.95	239	0.83	>0.2	22	0.85	264	0.96	>0.05
Dicots	1	10	1.00	55	1.11	>0.5	10	0.69	55	1.00	**<0.005**
	2	8	0.82	54	1.04	>0.2	8	0.70	54	1.00	**<0.0001**
	3	3	0.93	48	1.00	–	3	1.00	48	1.00	–

ACKNOWLEDGMENTS

Firstly, I thank Rob Gradstein for his trust, support and advice during the work on this thesis, but also for the opportunity to write up all the rest of the data we have gathered and to continue working at ECSF. Many thanks also to Michael Kessler for his advise, and to Doris Bär-Scheubel and Bernadette Tyson who probably have no clue how much I appreciate their help with bureaucratic issues.

Numerous colleagues helped invaluably with identifications, and I am much looking forward to making heavy use of their hundreds of reliable plant names soon: John Clark (STRI), Elvira Cotton (AAU), Tom Croat (MO), Stig Dalstrøm (SEL), Calaway Dodson (MO), Lorena Endara (QCA), Harald Kürschner (BGBM), Job Kuijt (University of Victoria), Marcus Lehnert (U. Göttingen), Carlyle Luer (SEL), James Luteyn (NY), Harry Luther (SEL), Jens Madsen (AAU), Juan Manuel Manzanares (QCNE), Guido Mathieu (GENT), Robbin Moran (NY), Benjamin Øllgaard (AAU), Susanne Renner (M), Harrie Sipman (BGBM), Alan Smith (UC), and Walter Till (WU). Zhoffre Aguirre (LOJA), Hugo Navarrete (QCA) and David Neill (MO, QCNE) are thanked for their friendly help with shipments, storage problems and for providing easy access to their herbariums. I am grateful to Helmut Dalitz and Mathias Ösker for generously (and repeatedly) lending their fancy and delicate hemispherical photography equipment. Florian Lauer and Meike Kühnlein helped analyse ArcGIS data, Christine Gehrig analysed microclimate data. Thanks to Konrad Fiedler for the statistics classes.

Financial support by the Deutscher Akademischer Austauschdienst (DAAD), Deutsche Forschungsgesellschaft (DFG) and IDEA WILD is most gratefully acknowledged. I mean it!

Big thanks to the staff at ECSF for making the place damn close to home: Tati and Rocío Aguirre and kids, Maria Feijoo (plus Angel, Jefferson, Robert and Carina), Polivio Ortega, Abrahan Pacheco, Janeth, Diana, and Pedro Paladines too. Thanks for memorable and extended company at ECSF (among many others) to Ruth Arias, Alfonso Arguero, Folkert Bauer, Jens Boy, Gunnar Brehm, Andrés Gerique, Sven Günter, los Javieres "Batata" y Clavijo, Jana Knuth, Nic Mandl, Steffen Matezki, Felix Matt, Christian Miersch, Rütger Rollenbeck (danke für viele gute Tips), Dirk Süssenbach, Alex and Beate Zimmermann, and especially to Glenda Mendieta, Pablo "Cerruchín" Ramirez (*), and Jürgen Homeier. Thanks also to Don Pepe Jacome and his staff for their hospitality at Bosque Protector Jerusalén, especially to Pato Moncayo, Jaime Puga and S. "Chino" Reyes.

Thanks also to Marcus Braun, Jorge Fabre, Rudy Gelis, Harold Greeney, David Hoddinott, Caleb Holtzer, Lou Jost, Andy MacLean, Elise Lockton, Nathan Muchhala, Barry Thomson, Fernando Vaca, Tom Walla, Richard White and Paula Quirros, Carla Bengtson, Federico Brown, Paola Carrera, Fabricio Guamán, Karim Ledesma, Maria, Monica, Tom and Mariella Quesenberry, Genoveva Rodríguez, Tom Shrives, Lori Swanson, Peter Wetherwax, and especially Alexandra Zach.

Special thanks to my parents, Lilo and Heiner Werner, for plentiful love, trust (apathy?) and opportunities granted. Abrazo de oso for my siblings Flip and Nina!

Die VDM Verlagsservicegesellschaft sucht für wissenschaftliche Verlage abgeschlossene und herausragende

Dissertationen, Habilitationen, Diplomarbeiten, Master Theses, Magisterarbeiten usw.

für die kostenlose Publikation als Fachbuch.

Sie verfügen über eine Arbeit, die hohen inhaltlichen und formalen Ansprüchen genügt, und haben Interesse an einer honorarvergüteten Publikation?

Dann senden Sie bitte erste Informationen über sich und Ihre Arbeit per Email an *info@vdm-vsg.de*.

Sie erhalten kurzfristig unser Feedback!

VDM Verlagsservicegesellschaft mbH
Dudweiler Landstr. 99
D - 66123 Saarbrücken
Telefon +49 681 3720 174
Fax +49 681 3720 1749

www.vdm-vsg.de

Die VDM Verlagsservicegesellschaft mbH vertritt

Printed by Books on Demand GmbH, Norderstedt / Germany